SUPERNOVA!

Also by Donald Goldsmith

Nemesis:
The Death-Star and Other Theories of Mass Extinction

Cosmic Horizons
(with Robert Wagoner)

The Evolving Universe

The Search for Life in the Universe
(with Tobias Owen)

From the Black Hole to the Infinite Universe
(with Donald Levy)

The Quest for Life in the Universe
(ed.)

Scientists Confront Velikovsky
(ed.)

SUPERNOVA!

THE EXPLODING STAR OF 1987

DONALD GOLDSMITH

ST. MARTIN'S PRESS
NEW YORK

Cover photo—Supernova 1987A is the bright star in the center of this photograph, with four "spikes" of light caused by the telescope. To the left of the supernova lies the enormous "Tarantula Nebula," a cloud of gas and dust lit from within by bright hot stars born within the past few million years. Both the supernova and the Tarantula Nebula are part of the Large Magellanic Cloud, a satellite galaxy of our own Milky Way. (Photograph copyright by European Southern Observatory (ESO), reproduced by permission of ESO.)

Design by ROBERT BULL DESIGN

Library of Congress Cataloging-in-Publication Data

Goldsmith, Donald.
Supernova!

1. Supernovae. I. Title.
QB843.S95G65 523.84'446 88-29861
ISBN 0-312-02647-1

First Edition

10 9 8 7 6 5 4 3 2 1

For Rachel—
Curiosity undiminished

CONTENTS

LIST OF ILLUSTRATIONS

ACKNOWLEDGMENTS

In writing this book I have been fortunate to receive the assistance of many astronomers active in fields of research that deal directly or indirectly with supernova explosions. I am grateful to all of them for their assistance, but would like to make clear that any errors in recounting the facts, or in describing the hypotheses, are my own responsibility. First and foremost, I would like to thank Bob Kirshner, Stan Woosley, and John Bahcall for their patience and generosity in dealing with my inquiries; they and other astronomers will understand that I have presented their scientific results without full credit to the others who have labored in the same fields, in order to present a more readable account. Ken Brecher has proven a fount of inspiration, scientific knowledge, and worldly wisdom. I am also grateful to George Blumenthal, Stu Bowyer, Chip Cohen, Lisa Ensman, Rich Epstein, Jim Felten, Alex Filippenko, George Field, Andy Fraknoi, Bob Garrison, Margaret Geller, Dan Gezari, Owen Gingerich, Paul Goldsmith, Woody Harrington, Dieter Hartmann, Minos Kafatos, Carolyn Kraus, Jon Lomberg, Steve Maran, Laurence Marschall, Dick McCray, Chris McKee, Toby Owen, Vahe Petrosian, Phil Pinto, Frank Shu, Steve Soter, Larry Sulak, Virginia Trimble, and Bob Wagoner for their help with this book. My editor, Michael Sagalyn, has once again demonstrated his sagacity in helping to organize this work and to make it more comprehensible. A special note of thanks goes to Supernova 1987A for exploding when and where it did, and for thus creating the flux of particles that led to this book.

INTRODUCTION

Close to 160,000 years ago, a star in the Large Magellanic Cloud, a galaxy close to our own Milky Way, burst itself in a majestic, violent death, an outstanding example of a supernova explosion. From this exploding star, light and other forms of radiation spread outward in all directions. Traveling six trillion miles each year, the radiation passed into dust-shrouded interstellar clouds, sped through the nearly empty reaches of the intergalactic medium, and radiated into other galaxies. Some of this radiation—about one part in a million trillion trillion of the total—reached our planet Earth, where it lit our skies as the brightest supernova explosion seen in nearly four centuries. By cosmic coincidence, the 160,000 years that passed on Earth between the explosion and the arrival of radiation from the supernova had included the evolutionary emergence of first the genus *Homo* and then the species *Homo sapiens,* creatures intelligent enough to manipulate and to contemplate their environment.

Supernova 1987A was special to astronomers not merely because it was visible—for, with thousands of relatively nearby galaxies to observe, astronomers now find ten or twenty supernovae each year—but also because it exploded in the closest galaxy to our own Milky Way. So close, in fact, that astronomers could do far more than study the supernova's visible light, as they could for supernovae thousands of times more distant than SN 1987A. The most important fact about the appearance of SN 1987A *in* 1987 was that astronomers could use newly developed and newly improved techniques to study the types of radiation and particles emitted from a supernova explosion in detail never before observed.

Had the supernova exploded at only 99.9 percent of its actual distance from Earth, the light from the supernova would have reached us during the presidency of John Quincy Adams. This would have been appropriate in view of the fact that Adams was the American president most interested in astronomy, but it would also have precluded the examination of this supernova by what we call modern techniques—neutrino detectors, satellite-borne telescopes, and computerized light-sensing devices to detect and to analyze the light captured by large ground-based telescopes. On the other hand, if the supernova's distance had been one-tenth of a percent *larger* than its actual distance, *Homo sapiens* would have been poised for much better observations—when the supernova's light arrived in about the year 2147.

In astronomy, you must take the universe as you find it—and as it finds you. However, it helps to be prepared, and the astronomers who had made calculation upon calculation of how stars explode were prepared: The observations of Supernova 1987A confirmed decades of their theoretical research. The great supernova of 1987, which actually exploded in about the year 158,000 B.C., set the following terrestrial records for exploding stars:

- For the first time, scientists detected particles called neutrinos from an exploding star, the definitive evidence that a star's core had collapsed. Until Supernova 1987A, only neutrinos from our sun, among all the objects in the cosmos, had been detected.
- For the first time, astronomers found that they had observed a star repeatedly *before* it became a supernova, so they had "before and after" photographs. These photographic records allowed them to discover more about which stars explode and why.
- For the first time in their observations of supernova explosions, astronomers could accurately observe the radiation called X rays and gamma rays that the supernova emitted. The observations helped to validate astronomers' theories of supernova explosions. These now triumphant theories call for the emission of gamma rays and X rays as the result of the decay of radioac-

tive nuclei, which heat the matter blown out from the explosion long after the heating caused by the initial blast wave from the explosion has died away.

- Supernova 1987A was the first stellar explosion to be observed anywhere within the "Local Group" of galaxies, the small galaxy cluster to which our Milky Way galaxy belongs, since the year 1885, and it was the first supernova bright enough to be seen easily with the naked eye since the year 1604.

This book tells the story of the great supernova of 1987, and describes the effects that this and other supernovae have had on the rest of the universe. Supernovae are not to be sneezed at: Among other things, they made us. Every molecule in our bodies contains atomic nuclei formed deep within stars that later exploded, spewing their products into interstellar space, where a later generation of stars and their planets—our solar system, for example—could form from the ruins of the old. To study supernovae is to appreciate the interrelatedness of the cosmos: Stars are born, shine, and die, and from their ashes new stars may be born, in a cosmic recycling that represents an important part of cosmic evolution.

In order to appreciate the story of supernovae, we must learn what happens to stars when they grow old. To do so, I invite the reader to follow the story of the discovery of Supernova 1987A, and then to learn about the observational and theoretical astronomers and astrophysicists who gathered and interpreted the wealth of data from the exploding star. The history of past supernovae deserves recounting, for it helps us to understand how supernovae interact with the universe, and how they produce such amazing quantities of energy from a single star. The latter chapters of this book present the basic facts about supernovae, as we understand them: What happens to stars as they age, and why some stars end their lives in explosion, while most fade away quietly; why some explosions produce neutron stars and pulsars, while others produce black holes; why supernovae shine in X rays and gamma rays; and why and how a tiny fraction of the titanic energy output from an exploding star emerges in the form of visible light.

I hope that the readers of this book will acquire a heightened awareness of our role in the universe. We are not simply tiny inhabitants on a cosmic speck of dust, but participants in the cycles of stellar birth and death that eventually yield the cosmic mulch that allows us to live.

D.G.
Berkeley, California
June 1989

SUPERNOVA!

1

DISCOVERY

FEBRUARY 23, 1987: The Chilean Andes in summer. Along the western Andean foothills, one of the great assemblages of telescopes on Earth includes three large astronomical observatories, spaced over 150 miles of the north–south foothills of the giant mountain range. The southernmost observatory, on the mountaintop called Cerro Tololo, is the Cerro Tololo International Observatory, the southern station of the National Optical Astronomy Observatories, a scientific organization that also operates the Kitt Peak National Observatory near Tucson, Arizona. A hundred miles to the north lies the European Southern Observatory (ESO), sponsored by the governments of West Germany, France, Italy, Holland, Belgium, Denmark, Sweden, and Switzerland. And on a nearby peak to the north of ESO lies the Las Campanas Observatory, operated by the Carnegie Institution of Washington. At the Las Campanas Observatory, the University of Toronto leases a subunit of the observatory site for its southern observing station.

All three observatories—Cerro Tololo, ESO, and Las Campanas—find themselves in Chile for the same reason: The Andes offer the finest views of the heavens from the southern hemisphere of Earth. Because of the concentration of land in the northern hemisphere, and because European culture first gave rise to modern astronomy, a northern-hemisphere bias dominated our observations of the skies for centuries. Observatories were built throughout Europe and then in the United States, where in 1920 the Mount Wilson Observatory became the site of the world's largest telescope, the 100-inch Hooker reflector, until the 200-inch Hale Telescope at Palomar Mountain, a hundred miles away, displaced it in 1949. But during the first seventy years of this century, no large telescope existed in the southern hemisphere that might provide a detailed view of the universe that we cannot see from the north.

For we live on a (nearly) spherical planet, which rotates every day around an axis that maintains a constant orientation in space. As a result, whenever you look at the sky from the northern hemisphere, vast areas of the sky never rise above your horizon. Instead, solid Earth always lies between you and those areas. (see Figure 1). The stars in the sky directly above the south pole of Earth, or in directions close to that pole on the sky, can never be seen from New York or Los Angeles: Such stars always lie so far to the south that you would have to look *through* the Earth to see them. As a result, a fraction of the entire sky that varies from fully one-half (for observations made at the north or south pole) through one-third (for observations in California) to one-eighth of the entire sky (for observations in Hawaii) remains unseen for observers in the northern hemisphere. Only an observatory at the equator could see the entire sky—and there are no great observatories at the equator, because no favorable mountain site for such an observatory can be exploited: Weather conditions near the equator, among other factors, make such exploitation unfeasible.

THE SOUTHERN SKIES AND THE CLOUDS OF MAGELLAN

Astronomers have not been slow to notice this gap in our ability to observe the universe, but until recently they have continued to devote most of their resources to northern observatories. One of the earliest examples of astronomical observation from southern latitudes occurred during the seventeenth century when Isaac Newton's contemporary Edmund Halley (after whom the comet is named) spent more than a year on the southern Atlantic island of St. Helena. From this tiny speck of land, which much later became famous as Napoleon's final place of exile, Halley carefully recorded the positions of celestial objects that could never be seen from England. After Halley's expedition astronomers put little effort into making southern-hemisphere observations for more than a century. Finally during the 1830s another British astronomer, John Herschel, spent four years at the Cape of

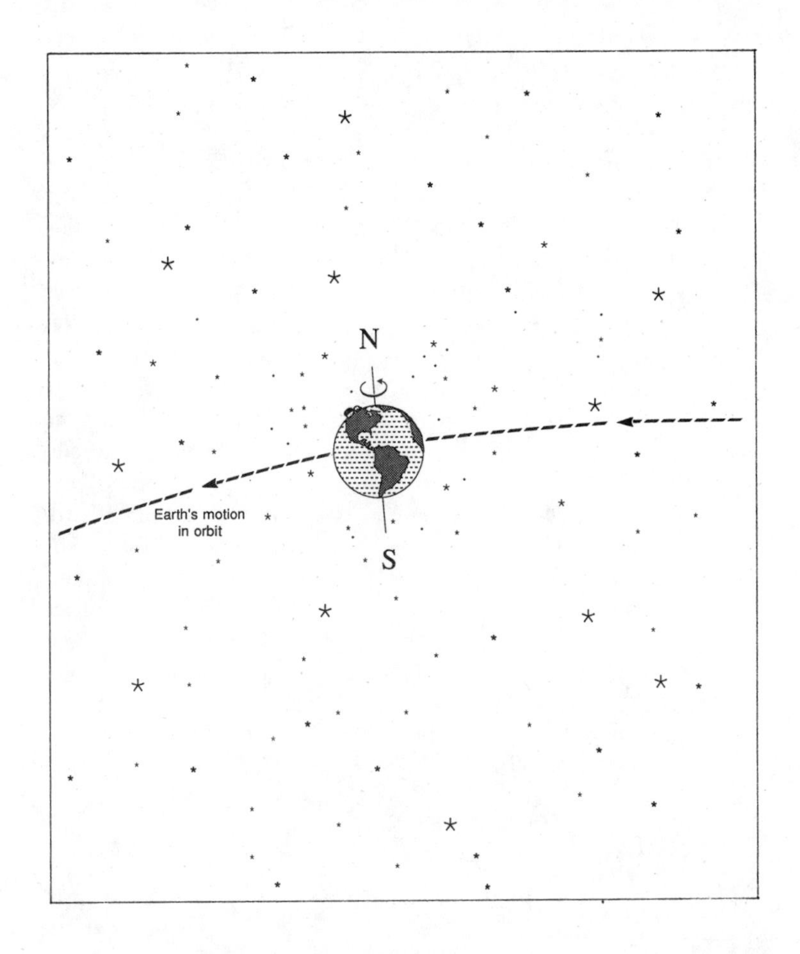

FIGURE 1. Because the Earth is round, and because the Earth's axis of rotation always points in the same direction as the Earth orbits the sun, observers on the Earth's northern hemisphere cannot see stars whose location on the sky places them close to the south celestial pole, the point on the sky directly above the Earth's south pole. (Drawing by Marjorie Baird Garlin)

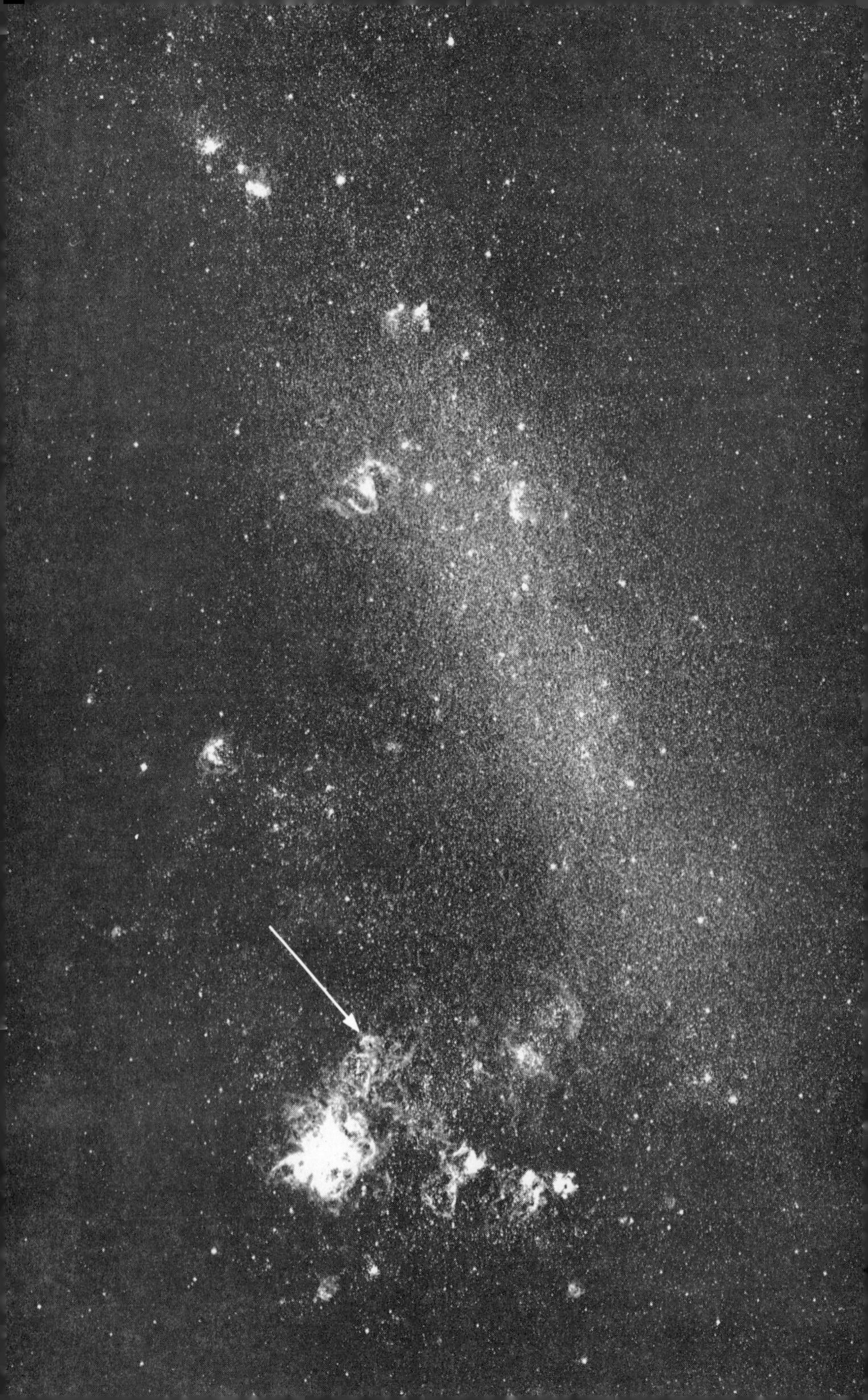

Good Hope in South Africa charting the southern skies in far greater detail than his predecessor had.

Herschel gave special attention to two "nebulae." These vast, diffuse clouds of light had first been called the "Cape Clouds" because sailors saw them as they neared the Cape of Good Hope at the tip of Africa. Then as now, even without a telescope to make them appear larger, these patches of light spread fuzzily over the sky, each as large as the full moon, though nowhere nearly as bright (Figure 2). Following Ferdinand Magellan's voyage around the world in 1519–1522, and the widespread publicizing of this voyage, the Cape Clouds were renamed the "Magellanic Clouds" or "Clouds of Magellan"; Magellan's sailors, like previous seafarers, had at times mistaken them for ordinary clouds, but eventually realized that they must be astronomical objects, not clouds at all. As another explorer wrote,

> [We] sawe manifestly twoo clowdes of reasonable bygnesse movynge abowt the place of the [south] pole continually now rysynge and now faulynge, so keepynge theyr continued course in circular movynge . . .

Ordinary atmospheric clouds drift irregularly, and often pass across the sky in an hour or less. In contrast, any celestial object has an apparent motion on the sky that simply results from the Earth's rotation: As the Earth rotates from west to east, we on Earth have the impression that we live on a stationary planet, while the "bowl of night"—the entire cosmos that we see—seems to turn from east to west. As the sky appears to turn, each celestial object maintains the same relative position with respect to all

---

FIGURE 2. The Large Magellanic Cloud, a satellite of our Milky Way, is an irregularly shaped galaxy that contains about 10 billion stars and has a distance of some 160,000 light-years from us. The galaxy curves downward to the left, embracing the Tarantula Nebula, a star birth region lit from within by young, hot stars. In this photograph, Supernova 1987A appears below and to the right (as we see it) of the Tarantula Nebula, the spiky cloud at the left center of the photograph. (National Optical Astronomy Observatories)

the others; thus the constellations essentially preserve their shape, night after night, year after year.

Because the Magellanic Clouds turned with the sky, and because they kept the same position in the sky with respect to the stars, astronomers and navigators recognized them as true celestial objects, not ordinary clouds that float in the air. John Herschel's telescopic studies of the Magellanic Clouds revealed within them a host of individual stars, star clusters, and smaller gas clouds, which we now know to be lit from within by groups of extremely young and luminous stars.

THE FABRIC OF THE UNIVERSE

The hierarchy of structure in the universe rests on stars, the fundamental luminous units of the cosmos. Each star is a sphere of gas, held together by its own gravitational force, so hot and dense at its center that nuclear fusion occurs, binding together the cores of atoms and steadily releasing new heat and light in a great flood that makes the stars shine. Around each star, relatively small objects called planets may—in the case of our solar system, certainly do—revolve, warmed by the light of their parent star and held in orbit by gravity. The planets' orbital motion causes them to fall "around," rather than into, their parent stars. Unfortunately, we cannot hope to see planets outside our solar system because the light they reflect in our direction is overwhelmed by the direct glow from the stars to which they belong. Impressive though these planets may be to us, they appear nearly insignificant on any scale that measures the sizes of or the masses contained in stars.

Stars cluster together in galaxies, each with many billions of stars. Our Milky Way, one of the largest (but hardly *the* largest) galaxies that we know, contains some four hundred billion stars, each basically like our own sun. The Milky Way has a flattened, disk-like shape, dozens of times larger in its two long dimensions than its one short one (see Figure 3). Our sun and its planets lie some 30,000 light-years from the center of the Milky Way, a distance about two billion times the distance from the Earth to the sun. Light traveling at 186,000 miles per second takes 8.3 minutes to leap the distance of 93 million miles from the sun to the Earth. But it takes 30,000 years to reach us from the center of the Milky

Way, so our view of the galactic center tells us not about how things are there now, but how they were when Cro-Magnon man marveled at the skies.

Beyond the Milky Way lies the realm of other galaxies, the "island universes" of a bygone generation of astronomers. Each of these galaxies lies so far from us that we see it as it was anywhere from a few hundred thousand to a few billion years ago (see Figure 4). The Large Magellanic Cloud and the Small Magellanic Cloud in fact are satellites of the Milky Way, moving in orbit around our galaxy as the result of mutual gravitational attraction. The Milky Way pulls on the Magellanic Clouds with its own gravity, and the Clouds pull back with theirs, but since the Milky Way has far more mass than either of the Clouds, they do most of the moving while our galaxy remains relatively still.

The Milky Way contains more than a thousand stars for every person in the United States. In contrast to the Milky Way's 400 billion stars, the Large Magellanic Cloud contains a mere 10 billion and the Small Magellanic Cloud about 5 billion stars. Billions of years ago, these three galaxies "pulled themselves together" by gravitational forces: They grew to be smaller and denser condensations within enormous clouds of gas and dust, each part of which attracted all the other parts by gravity. Eventually, within the contracting gas clouds that became galaxies, billions upon billions of individual stars likewise condensed from the material that formed the galaxy. These new stars were rela-

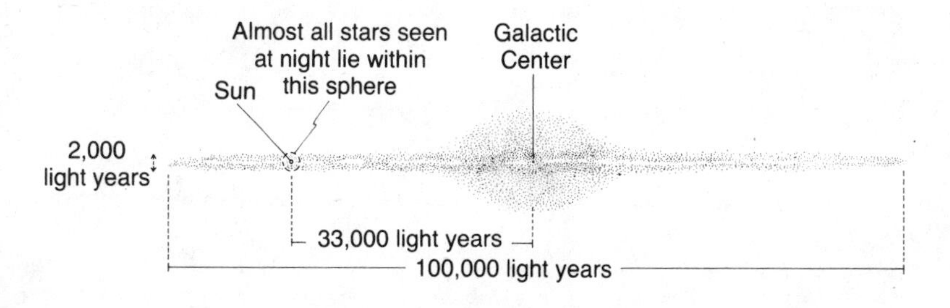

FIGURE 3. The Milky Way contains 400 billion stars in a highly flattened distribution, about 100,000 light-years across. The sun lies more than 30,000 light-years from the center of the galaxy. (Drawing by Marjorie Baird Garlin)

tively tiny condensations within condensations of far larger size. The Milky Way and the two Magellanic Clouds belong to what astronomers touchingly call the "Local Group" of galaxies. The Local Group spans about 2 million light-years, and contains not only the Milky Way and its two satellites, but also the Andromeda galaxy, a close cousin of the Milky Way, and *its* two main satellite galaxies, along with another baker's dozen of galaxies, all of them much smaller than the Milky Way and the Andromeda galaxy.

Astronomers' best estimates for the distances to the Magellanic Clouds, the closest galaxies to our Milky Way, are 160,000 light-years for the Large Magellanic Cloud and 170,000 light-years for the Small Magellanic Cloud. To put these distances in astronomical terms, we measure them in units of the "light-year," the distance light travels in one year, equal to nearly 6 trillion miles. Alpha Centauri, the closest star to the sun, is 4.4 light-years away; the stars you see at night are anywhere from 8 to 1,600 light-years distant; and the entire Milky Way galaxy has a diameter of roughly 100,000 light-years. The sun and its planets have an outlying location in the "suburbs" of the Milky Way, 30,000 light-years from its center. On a clear summer's night, when we catch a glimpse of the faint band of light spreading across the dark sky, which we call the "milky way," we can admire the host of stars near the central plane of our galaxy, whose light has traveled tens of thousands of years to provoke our wonder and understanding.

STARLIGHT MESSAGES

The enormous distances between objects in the universe implies that our knowledge of the universe must always be at least somewhat out-of-date. Astronomy is history, for we observe the universe not as it is but as it was, and at different times in the past,

---

FIGURE 4. This spiral galaxy in the constellation Pavo lies about 50 million light-years from the Milky Way. Like the Milky Way, this spiral galaxy is highly flattened, and shows its youngest, brightest stars concentrated into "spiral arms." (National Optical Astronomy Observatories)

depending on the objects that we observe. Only in our own solar system can we say that our understanding is up-to-date. Light from the sun, traveling at 186,000 miles per second, takes just over eight minutes to reach the Earth—eight minutes of grace in the (unlikely) event that our sun should disappear. Sunlight takes fifteen minutes to reach Mars, forty-five minutes to Jupiter, and an hour and twenty minutes to Saturn. Likewise, when we study the sunlight reflected from the planets, we see them as they were, not as they are. The few minutes' travel time in the solar system hardly matters, though it does prevent us from guiding spacecraft to a landing step-by-step; instead, we must pre-program the spacecraft to perform its own landing. But when you multiply the distances by a million, as we do when we look to the stars, you begin to encounter serious time delays. Every exploding star that has been recorded by the human race actually exploded thousands of years before (for the explosions in the Milky Way) or much longer ago than that (for the explosions in other galaxies).

WHAT MAKES A STAR EXPLODE?

Supernovae are exploding stars—the last, titanic gasps of particular types of stars that blow themselves up when they have run out of ways to continue their existence. All stars in the prime of life shine through the process of nuclear fusion, the melding together of the nuclei that form the centers of all atoms. This fusion changes one type of atomic nucleus into another. In the centers of stars, nuclear fusion changes hydrogen nuclei into helium, releasing light and heat. The fusion occurs only in the innermost core of a star, because only there is the star hot enough for nuclei to collide with sufficient violence to make them fuse together.

Toward the end of its life, the center of a star runs out of hydrogen "fuel." As it does so, the star contracts its core and consumes its remaining supply of fuel ever more rapidly. This makes the star release even more energy at its center, and some of this extra energy heats and expands the star's outer layers. As these layers expand, they cool. Hence the star, whose surface was formerly so hot that it glowed with a blue or yellow fire, turns red, the color of matter at temperatures of "only" a few thousand degrees.

Thus, during the later stages of its life, any star—our own sun included—will become an enormous, rarefied, "red giant," a star that has expanded to the point that its outer layers span nearly the distance from our sun to the Earth. Five billion years from now, the inhabitants of the Earth, if any, will watch in awe as our parent star swells tremendously, and its outermost layers threaten to engulf them, heating the Earth's surface to a temperature far above the boiling point of water. The swelling of the sun into a red giant conceals the heart of the star—its nuclear-fusing core—which will be contracting steadily, using the last of its nuclear "fuel."

Eventually, the fluffy outer layers of a red giant star will evaporate into space, leaving behind only the core itself. In most stars, including our future sun, this core stands revealed as a "white dwarf," a densely packed stellar cinder about the size of the Earth. The key point about white dwarfs is that nuclear fusion does not occur within them, so they shine only dimly from the energy that they stored as heat during their heyday as ordinary stars. As time passes, each white dwarf slowly cools and grows fainter, until invisibility becomes its fate. With certain important exceptions, all stars end their lives as white dwarfs, the shrunken, fading cores that once produced the heat and light that powered the star.

The exceptional stars are those that explode after they have become red giants and have lost their outer layers. Stars born with especially large masses—especially great amounts of matter—turn out to be unable to become white dwarfs. Instead, when such stars age, their cores undergo sudden collapse, falling in upon themselves within a small fraction of a second. This collapse triggers an outward explosion and creates a "supernova," a new star of exceptionally great luminosity. "Nova" means new in Latin, and in aging stars, a modest flare-up from time to time will produce a nova from time to time to catch the eyes of astronomers. But the novae (note the Latin plural) rank as nothing in comparison with the utter and final destruction of a star, so the prefix "super" quite naturally attaches itself to these exploding stars, which, during their few months of glory, can outshine a billion ordinary stars or 100,000 novae.

THE TWO TYPES OF SUPERNOVAE

Astronomers now have at their disposal several well-developed, though hardly complete, theories about what makes stars explode. These theories, as well as astronomers' observations of supernovae in far-distant galaxies, divide all supernovae explosions into two main types, affectionately known as Type I and Type II supernovae. According to this classification, stars that explode do so either because they cannot become white dwarfs and instead collapse (the Type II supernovae), or because they are *already* white dwarfs that have one final flare-up (the Type I supernovae).

TYPE I: WHITE DWARF STARS WITH ORBITING COMPANIONS

Type I supernovae are thought to arise from white dwarf stars. By itself, a white dwarf can produce no explosion. But if a nearby star—for example, a companion star in orbit with the white dwarf—deposits a sufficient amount of material onto the shrunken star, the white dwarf has a source of "new" fuel. The white dwarf's immense gravity causes this new material to fall onto the surface of the white dwarf, coating it with a nuclear manna that has the potential to release more energy through nuclear fusion. For a time, nothing happens as the infalling material accumulates on the white dwarf's surface, but finally a sort of flame of nuclear fusion ignites the entire star in a brief final moment of incandescent splendor that blows it completely apart.

TYPE II SUPERNOVAE: THE COLLAPSE OF STELLAR CORES

The second type of supernova, Type II, arises (in theory) as the final stage in the evolution of a massive star, a star with a mass equal to ten or twenty times the mass of our sun. These high-mass stars are relatively uncommon; in fact, our own sun has more mass than 90 percent of all stars. Astronomers have calculated how a star with ten or twenty times the sun's mass will contract its central core as it ages, raising the temperature there and fusing nuclei together to produce ever more complex types of nuclei. As the star heads for its rendezvous with destructive destiny, it desperately (if we may anthopomorphize for a moment) seeks additional ways to generate more heat through nuclear fusion, without which the star must collapse under its own gravity, the pull of each part of the star on all its other parts. As part of this quest for more heat, the star's core turns helium nuclei into carbon and oxygen, then carbon and oxygen nuclei into neon and magnesium, then neon and magnesium into silicon and sulfur, and finally silicon and sulfur into nickel and iron. But the struggle for more energy to heat the core must eventually fail, so the star's core collapses, unable to support itself against its own gravitation. The collapse stops when the core becomes an immensely dense "neutron star." The neutron star forms with an outward "bounce" that sends a fast-moving shock wave through the star's outer layers, blowing them into space in the violent, glowing event we call a supernova.

The very act by which a supernova attracts our attention—the outburst that makes it explode and emit enormous amounts of light—destroys our chance to study the pre-supernova star in detail. Hence our ability to distinguish between Type I and Type II supernovae rests not on our observations of a star's appearance before its explosion, but rather on astronomers' models of how stars explode, as well as from their observations of the actual explosions and the types of galaxies in which supernovae appear.

To discriminate between Type I and Type II explosions, astronomers spread the supernova's light into its various colors in a "spectrograph" and analyze it, color by color. This analysis reveals two important facts about the material ejected from the exploding star: the types of elements that it contains, and the speed at which the material is moving away from the star. From their observations, astronomers have found that Type II supernovae eject large amounts of hydrogen gas into space at enormous velocities, ten thousand miles per second and more. In contrast, Type I supernovae show no such fast-moving hydrogen, presumably because the stars that become Type I supernovae have already expelled their hydrogen-rich outer layers into space during their red-giant phases, and turned all the hydrogen in their cores into other elements through nuclear fusion.

To validate the distinctions that they make between Type I and Type II supernovae—Type I arising from white dwarfs with companions, Type II from massive stars that finally give up the ghost and collapse—astronomers would dearly love to observe a supernova *before* it explodes, if only to find out whether these theories of supernova explosions have merit. Although they have yet to do so as carefully as they would like, the supernova of 1987 brought astronomers closer to this goal than ever before possible. As usual, however, in doing so the supernova introduced as many mysteries as it resolved.

SERENDIPITOUS DISCOVERY: THE BEST KIND

On February 23, 1987 the Large Magellanic Cloud was easily visible from Las Campanas Observatory in Chile. Toward the west, the Atacama Desert, one of the driest spots on Earth, lay between the hills and the cold Pacific. At the observatory itself, the often long-immured, primarily male astronomers have a saying: "There's a woman behind every tree at Las Campanas." East of the treeless foothills that enfold the observatory, the high peaks of the Andes thrust toward the usually cloudless skies, silent and bare.

On that evening of February 23, the astronomers at Las Campanas began their routines, unaware that before the night was over, they would see the first light on Earth from a star that had exploded 160,000 years before, when our ancestors may have grappled with the notions of language. Most astronomers at Las Campanas, as at any other observatory, are short-term visitors, up on the mountain from their usual positions at universities or research institutions, lucky holders of the right to use one of the world's great telescopes for a few consecutive nights. But one of the smaller telescopes at such an observatory is often the domain of an astronomer permanently stationed—at least for several years' time—at the observatory, sometimes in connection with a long-term research project, sometimes as a sort of overall assistant to the visiting astronomers. At Las Campanas Observatory in the South American summer of 1987 there was one such astronomer, Ian Shelton.

Shelton is a Canadian from Manitoba, a man just past thirty who got the astronomy bug in his youth, built an observatory in his backyard (not the most common teenage diversion in Winnipeg), worked in a science museum, and pursued graduate studies in astronomy until he decided he'd rather try life in an observatory. By the late 1980s, Shelton was stationed for his second two-year stint as the resident observer of the University of

Toronto's Southern Observatory at Las Campanas, the possessor of fine views of the stars, of a small room in a stone dormitory next to the telescopes, and of a modest salary. Since Shelton rarely left the observatory, the salary was secondary: His interests lay in improving the telescopic equipment and in observing the skies. Shelton's character, not to mention the competing attractions in northern Chile, made him the sort of astronomer who on his night off decides to do some extra observing. Such astronomers are well placed to make serendipitous discoveries.

Shelton had requested the use of a small telescope, one with a 10-inch objective lens, that belonged to the Las Campanas Observatory. The resident scientist and administrator, William Kunkel, had assented. The 10-inch telescope is of the type called an "astrograph"—a telescope with a relatively wide field of view, useful for taking survey pictures of the sky. This astrograph had been built before the Second World War, and had given years of excellent service at the Mount Wilson Observatory outside Pasadena, California. During the early 1950s, the telescope was sent to South Africa, where a young astronomer from the University of Michigan, Karl Henize, who would later become a United States astronaut, had used it to survey the southern skies, discovering many remnants of exploding stars and including many a photograph of the Large Magellanic Cloud in his work. Still later, after smog and city lights had made Mount Wilson less attractive to astronomers, the telescope was sent to Chile. Shelton, who had used the astrograph to photograph Halley's comet the year before, wanted to use it to take a three-hour-long wide-field photograph of the Large Magellanic Cloud, in order to test how well the instrument's guiding system could follow the Cloud's motion across the sky as the Earth turned. Shelton had taken a similar photograph on the previous night, but it had not turned out as well as he had hoped.

On that night of February 23, Shelton once more pushed the sheet-metal roof out of the way, opened the slide that exposed the photographic plate, and spent the next three hours peering into the guide telescope, making fine adjustments to keep the telescope pointing directly at the Milky Way's large satellite. With the exposure complete, Shelton went to get his coat, for the wind was rising, and then began another exposure. Soon afterward, the wind blew the roof shut, luckily leaving the telescope unharmed. By now, in the predawn hours of February 24, Shelton judged that his night of observing was complete, and went off to attempt to sleep until noon, as astronomers routinely do without shame. But he decided that first he would develop his photograph of the Large Magellanic Cloud, since the darkroom occupied part of the house in which he lived.

To an astronomer who knows the Large Magellanic Cloud as well as Ian Shelton does, a large bright spot on this photographic plate was enough to make his heart rise. Such a spot must be either a flaw in the plate or a relatively bright star, and Shelton knew that no star as bright as his image existed in that region of the sky (Figure 5). Turning over the possibilities in his mind for a number of minutes led Shelton back to the real world: He went out into the windy night on what by now was the morning of February 24 and looked at the Large Magellanic Cloud. And there was a new star, shining near a part of the galaxy called the

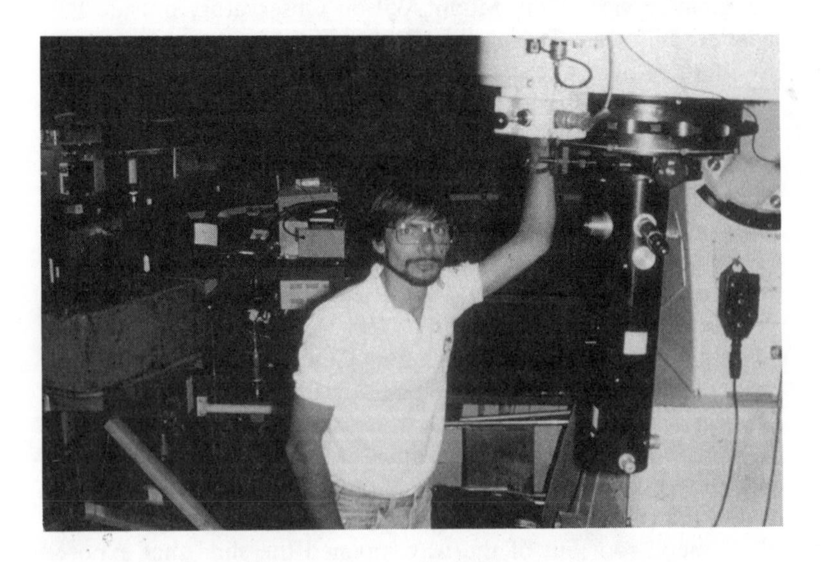

FIGURE 5. Ian Shelton, of the University of Toronto, discovered Supernova 1987A on February 23, 1987, by photographing the Large Magellanic Cloud and noting that a new star had appeared since the previous night. (University of Toronto)

"Tarantula Nebula," a cloud of gas and dust. Since Shelton had photographed the Large Magellanic Cloud on the previous night, he knew for certain that he was observing a new star.

Shelton walked uphill to the dome of the much larger 40-inch reflecting telescope, operated that night by another Canadian, Barry Madore from the University of Toronto, by Robert Jedrzejewski, a young astronomer from the Carnegie Observatories, and by Oscar Duhalde, one of the technicians employed to help operate the Las Campanas Observatory. Shelton inquired what they would think of a "fifth-magnitude object" (a star bright enough to be seen without a telescope, but not one of the brightest stars in the sky) in the "LMC," the Large Magellanic Cloud. "You're kidding," Madore said, while Duhalde nodded in agreement with Shelton, possibly ruing his missed opportunity—for he then recalled that two hours before, during his coffee break, he had noticed that the Tarantula Nebula seemed unusually bright. The star Madore had failed to notice, Supernova 1987A, is now sometimes referred to as Supernova Shelton–Duhalde–Jones (Figure 6). This commemorates Duhalde, the first person to *see* the supernova, as well as Shelton, the first person to identify the object *as* a supernova, and Jones.

And who is Jones? Albert Jones is an enthusiastic and dedicated amateur astronomer in New Zealand, who also happened to be observing the Large Magellanic Cloud that night, as part of his almost-nightly observational routine. Quite familiar with the appearance of the Magellanic Clouds, Jones saw the supernova and recognized it as a new star only a few hours after Ian Shelton did. (It was still dusk then in New Zealand, although in Chile daylight had already broken.) Jones's detection preceded that of a professional astronomer, Robert McNaught, who was observing the Large Magellanic Cloud from the Siding Spring Observatory in Australia. McNaught had taken a photograph of the Large Magellanic Cloud on the previous night, which he had developed but had not examined on the night of February 23. Since Australia lies to the west of New Zealand, McNaught did not see the new star on that night until an hour and a half after Jones did, though he was in fact the first person to *photograph* the supernova.

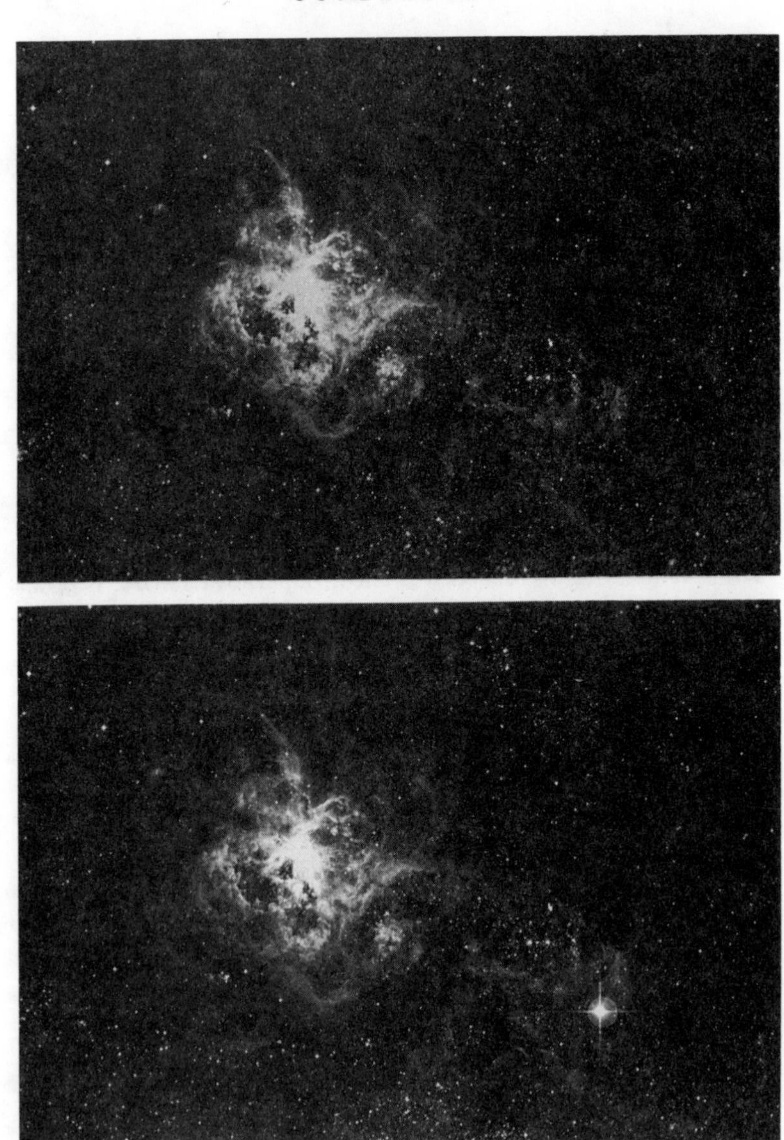

FIGURE 6. These photographs show the region surrounding the Tarantula Nebula before (in 1982) and a few days after (February 27, 1987) the supernova appeared. (Royal Observatory, Edinburgh)

THE *SILVER BLAZE* EFFECT

Jones and McNaught indeed made early observations of the supernova, but in a perverse twist completely familiar to scientists, their most important result consists of what Jones did *not* see, rather than what he *did* see. What Jones and McNaught *saw* provided confirmation of Ian Shelton's detection of the supernova, but Jones's *failure* to observe anything on the *previous* night provided a key upper limit on the maximum brightness that the supernova might have had on that night.

Why is the supernova's maximum possible brightness on the night before its discovery so important? The significance lies in analysis of the elusive particles called "neutrinos" that were produced in the supernova explosion, which we shall discuss in Chapter 4. Using specialized detectors buried deep underground, four different teams of observers recorded a "pulse" of neutrinos—far more than usual—at two different times: Three of the detectors registered a burst of neutrinos about four hours after the other one did.

Astronomers agreed that one of these two times of "detection" must represent an actual observation, while the other must be some sort of observational fluke. One of the possible neutrino-detection times contradicted the fact that Jones failed to observe the supernova on the previous night, for the theory of supernova explosions implies that if the neutrinos had arrived at the earlier time, Jones should have been able to see the new star on the night that he failed to do so. The second neutrino-detection time, four and a half hours later, was just enough later that it could fit with Jones's failure to observe the supernova. Scientists sometimes refer to a failure to see something as the "Silver Blaze effect," named after Arthur Conan Doyle's Sherlock Holmes story, "Silver Blaze." In this tale, the horse Silver Blaze turns out to be the "murderer," and Sherlock Holmes realizes that whoever removed the horse from his stall must have been known to the dog that guarded the stables:

> "Is there any other point to which you would wish to draw my attention?"
>
> "To the curious incident of the dog in the night-time."
>
> "The dog did nothing in the night-time."

"That was the curious incident," remarked Sherlock Holmes.

Jones's "Silver Blaze" observation of the Large Magellanic Cloud *without* the supernova proved crucial in resolving the issue of which "detection" of neutrinos from the supernova could be regarded as correct. Ian Shelton had also photographed the Large Magellanic Cloud on the night of February 22–23, and had recorded no new star. However, Jones and McNaught were observing many hours later, since they had to wait for night to fall across the Pacific. As a result, Jones's observation in New Zealand set the crucial "upper limit," the maximum possible brightness that the supernova could have had at the time that Jones did *not* see it, for any greater brightness would have meant that he *did* see it. Jones, scanning the Large Magellanic Cloud with his telescope, simply saw nothing, but when McNaught examined the photographic plate that he had exposed in Australia on the prediscovery night, using a more sensitive telescope than Jones had, there was the supernova, still quite faint but about to grow much brighter! Had McNaught developed his plate and studied it in detail immediately after exposing it, the previous pages might have described his astronomical background, not Ian Shelton's. Of such luck are discoveries made, but Jones and McNaught deserve a salute nonetheless.

HOW THEY BROUGHT THE GOOD NEWS FROM LAS CAMPANAS TO CAMBRIDGE

Ian Shelton's news had a slow start to civilization, but a rapid run around the world once it tied into the network that astronomers use to signal important discoveries. Because the radiophone from Las Campanas to the observatory headquarters in the city of La Serena failed to work, Oscar Duhalde and Angel Guerra, another observatory assistant, drove to La Serena and sent a telex to Cambridge, Massachusetts, where the Central Bureau for Astronomical Telegrams was established two decades ago. The telex reached Cambridge just before a telephone call from New Zealand that brought news of Jones's observation. The news promptly went out to observatories and astronomical institutes the world over: Ian Shelton reported a new object in the Large Magellanic Cloud, "a mag 5 object, ostensibly a supernova."

The Central Bureau had reported supernovae many times before, but never in a galaxy so close to us. When the telegram reached the world's observatories, one of two reactions ensued. Those too far north to observe the Large Magellanic Cloud sighed, while those to the south sprang into action. In Chile, New Zealand, Australia, and South Africa, no observer wanted to miss the explosive event of a lifetime. Regular observing programs were shelved, telescopes were trained on the Large Magellanic Cloud, and the first systematic observations of the rise and fall of the supernova's light began.

THE STAR THAT EXPLODED: SK −69° 202

Within hours, astronomers had studied their old photographic plates and had located the most likely candidate for the star that exploded (Figure 6). This luminous blue star had the name Sk −69° 202. "Sk" from the fact that it appeared in a list of bright blue stars compiled several years before by Nicholas Sanduleak ("Sk") of Case Western University: −69 for its declination (an astronomical coordinate), which is 69 degrees south of the celestial equator; 202 for its position on the list of stars at that declination. Not surprisingly, the star soon became known as "202," like a locomotive in the glory days of railroading.

Sk −69° 202 had been photographed for years without showing any particular claim to astronomers' attention. Apparently the star had gone bump in the night without much warning that an explosion was imminent. As explained in Chapter 6, one of the great mysteries of Supernova 1987A arose from the fact that 202, the star that blew up, was not the sort of star that astronomers believed ripe for explosion. This gave them a few weeks of worry, until theoretically minded astronomers showed that indeed a star like 202 could be a candidate for supernova explosion.

Suitably reassured, astronomers used their experience of previous supernova observations to predict that SN1987A would reach a maximum luminosity sufficient to give it the apparent brightness of the brightest stars in the sky, a southern hemisphere marvel. Instead, SN 1987A peaked at a brightness that made it merely one among hundreds of stars, so far as apparent brightness goes, never brighter than one of the stars in the Little Dipper. The supernova could be seen without a telescope, but you had to be rather astronomically adept (as well as located in or near the southern hemisphere) to observe it, even in May 1987, when the supernova reached its peak luminosity. To put things in perspective, the supernova became noticeably brighter than Halley's Comet (another well-advertised object best seen from the southern hemisphere) had been during the previous year, but as with Halley's Comet, you did not wander outside and ask, "What's that amazing object in the sky?"

Brightest of all or merely one among hundreds, SN 1987A was a big hit with astronomers. To see why this was so, we must make a small excursion into the science of astronomy, in order to understand how we learn about the universe through the hidden messages of starlight. Afterward, we can take a closer look at the astronomers who seek to unveil the mysteries of the universe through observation and—equally important—through calculation. These two types of astronomers, observers and theoreticians, have grown somewhat apart during the past few generations: One sort travels to distant mountaintops, or to radio observatories placed as far from civilization's interference as possible, while the other stays at home deep in thought, ready to use giant computers to explore theories of the universe. Among the greatest virtues of Supernova 1987A lies the fact that it demonstrates what some astronomers concede only ruefully: We need both types.

2

THE HIDDEN MESSAGES OF STARLIGHT

As Supernova 1987A once again demonstrated, nearly everything that we know about the universe depends on what we *see* when we look into the vast and glorious cosmos that surrounds us. Until a few centuries ago, a "look" at the universe was just that: Either you relied on your own eyes, or you found the person with the best pair of eyes, and learned what he or she could see in the clear night sky. Eventually, during the early years of the seventeenth century, scientists learned how to combine two curved lenses to make a telescope, an instrument that makes objects seem closer to the viewer. With telescopes we could collect faint starlight, could read its messages far more clearly than before, and could discover, as Galileo did, the mountains on the moon, the satellites of Jupiter, and the myriad stars that form the band of light called the "milky way."

However, the first telescopes produced such fuzzy images that long training was required to understand what the user was seeing. Gradual improvements in telescopes, and in the instruments used to detect and to analyze the light that our telescopes capture, have provided us with far clearer pictures of the cosmos than our ancestors could command. With modern telescopes we can see objects 100 million times fainter than the faintest stars visible to the unaided eye. Impressive though this achievement may be, it pales in comparison with the real breakthrough of modern astronomy: We have learned to see the cosmos in waves completely invisible to the human eyes.

As Supernova 1987A abundantly demonstrated, the messages

of starlight arrive not only in what we call "light" but also in other types of waves, similar to light but completely hidden to human perception. Today, thanks to our technological achievements, we can detect with human-made sensors what human bodies cannot perceive, a host of waves that reach the Earth, invisible to our eyes but accessible to our instruments.

THE NATURE OF LIGHT

Light waves are one particular type of "electromagnetic waves." The term "electromagnetic waves" reflects the fact that they can be produced through the rapid movement either of electrically charged particles or of particles that possess magnetism. If electromagnetic waves pass by an electrically charged particle, they will make it oscillate back and forth, as if the particle were being rocked by a wave like the wave on a pond. But what are these waves? What forms the light that we see?

This question baffled Aristotle, confused Isaac Newton (even as he made important discoveries about light), and remained a mystery, despite improvements in our scientific understanding, until Albert Einstein and Niels Bohr spent a few years with it during the early years of this century. Today, scientists have no hesitation in providing the answer: Light waves are waves that are analogous to the waves in water, but that require nothing—no water, no air, no matter of any sort—to travel through the universe!

PHOTONS: BOTH WAVES *AND* PARTICLES

The ability of light waves to pass through empty space lay at the heart of scientists' difficulty in understanding that waves need *something* to wave in. Outside of such a "medium," something for the waves to ripple through, we would *expect* to find not waves, but *particles,* objects that exist independently of anything else.

Bohr and Einstein eventually saw the solution to the mystery of light. Light waves are both waves *and* particles! On the one hand, since light and other electromagnetic waves move freely through empty space, they resemble particles. We call the particles that form light waves "photons." Thus, *electromagnetic waves are streams of photons.* We can imagine that every source of light shoots out photons that always travel through empty space at the same speed, the speed of light, 186,000 miles per second. But light and other electromagnetic waves possess a crucial property of waves: If you hold an electrically charged particle as an electromagnetic wave passes by, it will oscillate up and down in a cyclical manner, just as a toy boat on a pond will bob up and down when a water wave passes, rippling outward from a stone dropped into the pond.

The trick to understanding electromagnetic waves is to think of photons as *particles with wavelike properties.* Imagine each photon as a sort of evanescent tadpole, waving its tail up and down as it travels at the speed of light. The number of times that the tadpole waves its tail up and down (one "cycle" of oscillation) in a second is the *frequency* of the wave. Different photons have different frequencies, all of which are measured as a number of cycles per second. The distance that the tadpole travels while it waves its tail up and down once is the photon's *wavelength.* If we were studying the ripples on a pond, the "wavelength" would provide the distance between two neighboring wave crests. For the tadpoles that represent photons, the wavelength gives the distance in which the tadpole performs a back-and-forth oscillation.

Because the frequency measures the *number* of oscillations per second, photons with larger frequencies will require *less* time for each oscillation than the photons with smaller frequencies do. Therefore, since all photons travel at the same speed—the speed of light—the photons with larger frequencies travel *lesser* distances as they undergo a single oscillation than do the photons with smaller frequencies. Hence the photons with *larger* frequencies have *shorter* wavelengths than the photons with smaller frequencies. Conversely, the photons with smaller frequencies—slower vibration of their oscillating tails—travel farther as they perform a complete oscillation, and therefore have longer wavelengths, than do the photons with larger frequencies.

THE COLORS OF LIGHT

Frequency and wavelength are important because they distinguish one type of photon from another. Our eyes and brains perceive the differences between the frequencies and wavelengths as differences in *color* of the light that we see. Red light has the longest wavelength and the smallest frequency, violet the shortest wavelength and the largest frequency. All the other colors—orange, yellow, green, blue, indigo, and a thousand intermediate shadings—have frequencies and wavelengths that lie between the red (long-wavelength) and violet (short-wavelength) bounds of the light that we can see.

PHOTONS AS CARRIERS OF ENERGY

As a photon travels through space, it carries energy with it. The amount of energy that the photon carries is directly proportional to the photon's frequency, and inversely proportional to its wavelength. Since violet photons each have about twice the frequency and half the wavelength of red photons, they each have about twice the energy of a red photon. We can imagine the universe as filled with countless tiny packages of energy, each of them traveling at the speed of light—the photons that are emitted from countless sources of photons in all directions.

If you arrange to capture some of these photons—for example, by putting your eye in their path—the photons that you capture will deposit their energy in your eyeball. This act transfers the energy in the photons to the molecules in your eye, and the transfer of energy makes the photons disappear: A photon without energy has no claim to existence, for its energy is its essence. But the transfer of energy allows us to see the world around us.

When a photon enters a human eye, it passes though the lens and cornea, to encounter the retina at the back of the eyeball. Upon impact on the retina, the photon's energy produces a chemical change in molecules called "rhodopsin," which are found in the retina and nowhere else in our bodies. The chemical change causes an electrical impulse to pass from the retina to the brain, which interprets the impulse as part of a scene that our brain creates for us to admire, deep inside the skull. The world is more complex than we may imagine, and its complexity begins with the fact that we never see the "real" universe. Instead, our brains reconstruct a scene from tiny electrical currents from the retina after photons from the universe around us have carried their energy into our eyes, providing a view—but a limited one—

of whatever surrounds us. The limit on our view arises from the fact that our retina can detect only photons within a certain range of energies, frequencies, and wavelengths, the range that corresponds to what we call "light."

THE INVISIBLE UNIVERSE

Beyond the boundaries of light—at wavelengths longer than those of red light, or shorter than those of violet light—lies an invisible universe, a universe inaccessible to human knowledge until a few generations ago. The actual universe, as opposed to the universe that we *see,* knows nothing of the bounds of light waves, which arise from the details of human perception. Instead, the universe glows not only with the electromagnetic waves that we call light, but also with other types of electromagnetic waves, waves that the human eye cannot see, revealed to us today because we have mastered the technology needed to detect these other types of electromagnetic waves.

SLICING THE CAKE OF ELECTROMAGNETIC WAVES

We can picture "light," the range of photon wavelengths and frequencies that our eyes can detect, as a tiny slice of reality, a single layer in the cake of electromagnetic waves (Figure 7). This cake has an infinite number of layers, because photon frequencies and wavelengths can theoretically grow infinitely large; in practice, photons with enormous wavelengths don't travel far before being absorbed by electrons in interstellar space, and photons with enormous frequencies are only rarely produced. The astronomical action usually lies in the middling layers, the ones on either side of the visible-light layer.

INFRARED AND RADIO

"Infrared" photons have wavelengths slightly longer (hence frequencies somewhat less) than those of visible light. On Earth, infrared waves fill our environment with electromagnetic waves that our eyes cannot detect. All objects not at a temperature of absolute zero emit photons naturally, a process which we shall examine in detail when we consider how stars produce their light.

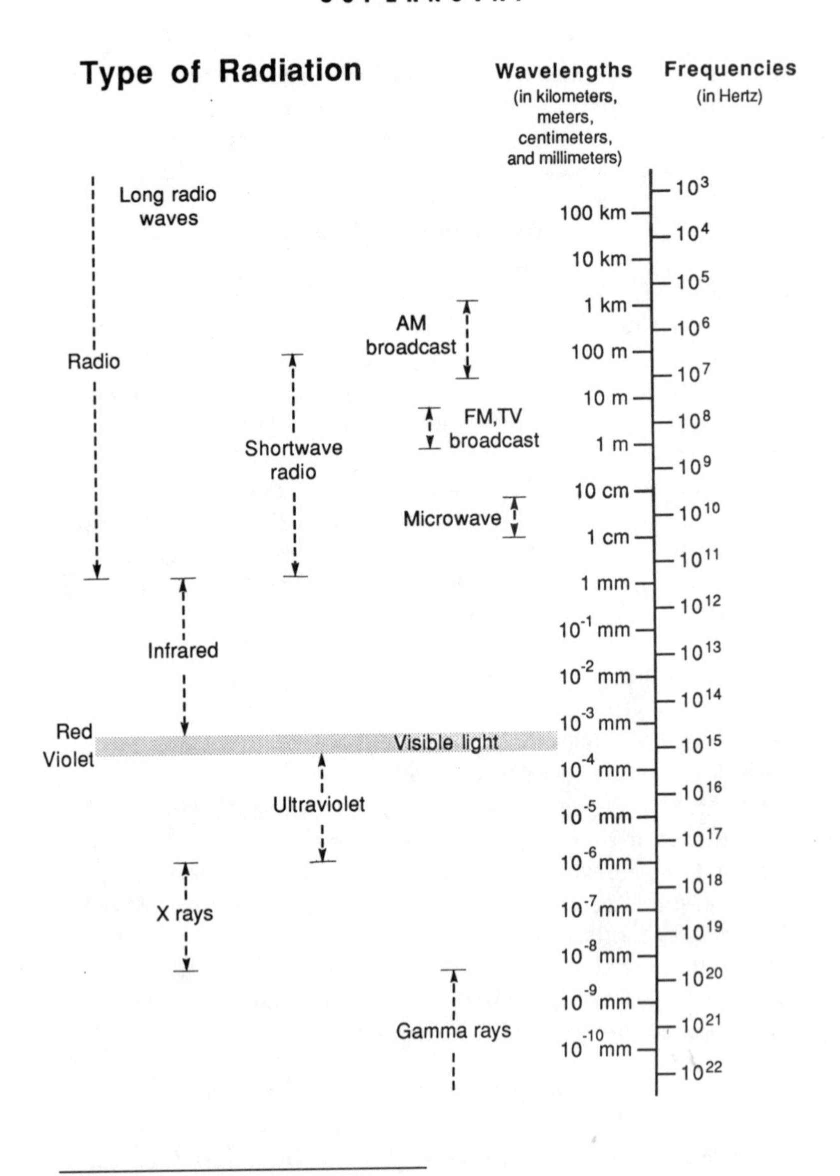

FIGURE 7. The "spectrum" of electromagnetic radiation extends over all possible wavelengths and frequencies (measured in cycles per second, or hertz). Visible light forms only a tiny portion of the total spectrum of this radiation. (Drawing by Marjorie Baird Garlin)

Objects at temperatures like those on Earth emit relatively large numbers of infrared photons. If we had "infrared eyes," we could see objects even in what we call pitch darkness, since objects on Earth continue to emit infrared photons whether or not visible light shines on them. The military forces of the world, relying on this fact, have developed sensitive infrared detectors that achieve exactly this end.

Below infrared in frequency, at still longer wavelengths, lies the domain of radio waves, sometimes subdivided into "microwaves" (shorter-wavelength radio waves) and "ordinary" (longer-wavelength) radio photons. Microwaves have found an important application in cooking: We arrange for a microwave oven to produce large amounts of microwaves, which deposit their energy in water molecules, because the ovens emit frequencies that are matched to the wavelengths at which water molecules vibrate. Because water molecules are abundant in the food that we place in the oven, microwave ovens "work" because their microwave photons make the water molecules in the food jiggle back and forth, heating the food and, subject to the skill of the cook, producing a tasty meal.

Though longer-wavelength radio waves are useless for cooking, they have a still more important property, which likewise arises from their long wavelengths: They pass through what we call solid walls without difficulty. This fact has made radio waves the dominant means of diffusing information widely and cheaply on Earth. We simply broadcast radio photons from a central transmitter, having coded these waves so that when they interact with a detector system, they create sound or, in the case of television, sound plus pictures. In theory, we could use any other type of waves, such as infrared or visible light, for the same purpose, but these types of photons would be far more easily blocked than radio waves. Hence radio waves have won the day, and we receive much of our information about human activities via radio photons.

ULTRAVIOLET, X RAYS, AND GAMMA RAYS

If visible light waves are mild ripples, then ultraviolet waves are stronger waves, X rays are tremendous breakers, and gamma rays more energetic still. We rank these types of electromagnetic waves not by the photons' wavelengths, which are amazingly small, but instead by the *energy* that each photon carries. Gamma rays are the highest-energy electromagnetic waves, and each gamma-ray photon has at least a million times the energy of the photons that form light waves. X-ray and ultraviolet photons have less energy than gamma rays, but still far more than visible light. Hence, when you look for these photons in the universe, beware! The high energy in ultraviolet, X-ray, and gamma-ray photons can destroy human tissue.

THE EFFECT OF THE EARTH'S ATMOSPHERE ON PHOTONS

Luckily for us, the Earth's atmosphere blocks out all these short-wavelength, high-frequency photons, shielding us from the potentially deadly effects of ultraviolet, X rays, and gamma rays. Life on Earth has evolved beneath a protective blanket of air; without it, we would hardly last a day, for the influx of high-energy photons (primarily from the sun) would prove lethal. For the time being, we find almost no high-energy photons in our local environment, and must go beyond the atmosphere if we hope to observe what sources of ultraviolet, X rays, and gamma rays exist in the universe. Most stars with surface temperatures like our sun's, or higher, emit copious ultraviolet radiation, but no normal star emits more than a tiny fraction of its energy output as gamma rays or X rays.

The boon of an atmospheric shield works a hardship on astronomers. Those who seek to observe the universe in gamma rays, X rays, and ultraviolet photons face a difficult problem in their work: Almost none of the types of photons they study can reach the ground. Astronomers are not fools, of course, and they consistently pause to pay tribute to our atmosphere even as they lament their difficulties in observing the cosmos. The rest of us may worry less about astronomers' problems, and simply count ourselves fortunate to have an atmosphere that protects us from the electromagnetic waves that would otherwise destroy us. The most important part of this protection consists of the shield against ultraviolet waves, which the sun produces in great amounts. This atmospheric shield arises from one of the most famous and least understood of molecules, ozone.

A molecule consists of two or more atoms held together in a

single unit. Perhaps the most familiar molecule is water, whose molecules are each made of two hydrogen atoms and one oxygen atom (H_2O). The air that we breathe contains mainly oxygen molecules (oxygen atoms linked together in pairs, hence described as O_2) and nitrogen molecules (pairs of nitrogen atoms, written N_2). These molecules provide some protection against streams of ultraviolet radiation from the sun and (much less important) other astronomical objects. Although some ultraviolet photons will disappear when they strike oxygen and nitrogen molecules, most of the ultraviolet would penetrate the atmosphere and strike the Earth's surface, were it not for ozone.

OZONE: OUR SHIELD AGAINST ULTRAVIOLET WAVES

Ozone is a relatively rare type of molecule, formed when oxygen atoms link together in triplets (O_3) rather than in pairs. Not much attention would be paid to ozone, were it not for the fact that ozone molecules prove remarkably efficient at absorbing ultraviolet photons. Because of this efficiency, even a relatively tiny number of ozone molecules, high in the stratosphere, suffice to remove almost *all* the ultraviolet radiation that streams toward the Earth. Ozone molecules are formed from oxygen atoms that are broken apart by high-energy ultraviolet photons: Some of the individual oxygen atoms recombine to form oxygen triplets (ozone) rather than doublets (oxygen molecules).

HUMAN ASSAULT ON THE EARTH'S OZONE SHIELD

The past few years have made it clear that human activity has gradually depleted the amount of ozone in the upper atmosphere. This depletion of ozone at high altitudes arises from our release of "chlorofluorocarbon" molecules, mostly from spray-can propellants and from refrigerator coolants (some of which inevitably leak, over time, from the cooling mechanisms). The chlorofluorocarbon molecules eventually rise into the stratosphere, where they combine with startling efficiency with ozone: A few chlorofluorocarbon molecules can "eat" a host of much smaller ozone molecules. Since we have no way to remove chlorofluorocarbons from the stratosphere, and since we depend on stratospheric ozone for our survival, it seems odd that we continue to use chlorofluorocarbons in refrigerators, the more so as the depletion of ozone, at least over the South Pole (the "polar ozone hole") has been confirmed beyond much scientific doubt. Although the United States has banned the use of chlorofluorocarbons in spray cans, other countries have not done so. We may yet owe our demise to an overconsumption of hair spray and refrigerators—an end predicted by comedians long ago.

TO SEE THE UNIVERSE, RISE ABOVE THE ATMOSPHERE

Meanwhile, we continue to live under the protection of a highly efficient ozone shield, which blocks nearly all ultraviolet radiation. Other types of atoms and molecules in the atmosphere screen out all X rays and gamma rays, the photons of the highest energies, and still other types of atoms and molecules, most notably water vapor and carbon dioxide, remove nearly all of the incoming infrared radiation. From an astronomical viewpoint, we live in a fish bowl too opaque to reveal the universe. If we seek to observe the cosmos in infrared, for example, we must carry our equipment at least to the summits of high mountains, or (still better) as high as aircraft can fly, in order to place ourselves above most of the water vapor.

Because of the atmospheric blockage of infrared waves, the Mauna Kea Observatory, nearly 14,000 feet above sea level on Hawaii's tallest mountain, has become the world's prime site for infrared observations of the cosmos. Even at Mauna Kea, however, some of the infrared waves cannot penetrate through the atmosphere to reach the observatory.

In order to observe the universe at those infrared wavelengths, we must fly telescopes eight or nine miles high in the Kuiper Airborne Observatory, or KAO. The KAO resembles a Boeing 707 aircraft, but has a large hatch in its roof that can be opened once the aircraft reaches its maximum altitude, close to 45,000 feet. Then the astronomers and technicians, wearing oxygen masks, attempt to detect infrared emission of a particular wavelength and frequency. By raising themselves not to the 14,000 feet of Mauna Kea, but all the way to 45,000 feet, the astronomers rise above most—still not all!—of the molecules in the atmosphere that absorb infrared waves, and thereby open a metaphorical "infrared window" of observation. The KAO flies regularly from NASA's installation at Moffett Field, near San Francisco, far out over the Pacific Ocean and back again, carrying the scientists and the equipment to make infrared observations of the cosmos far more easily and inexpensively than a satellite-borne observatory would. But for certain infrared wavelengths and frequencies, you still need a satellite—or a willingness to forego the observations.

If we seek to observe ultraviolet waves, X rays, and gamma rays from distant objects, we find that the absorbing atoms and molecules are too high for even the KAO's flight to 45,000 feet to help. To put the matter plainly, with the exception of a few infrared frequencies that can penetrate our atmosphere, to observe the universe in photons other than those of visible light and radio requires satellites that orbit entirely beyond the Earth's atmosphere. This effort, the most important achievement of modern astronomy, remains in its infancy, but has already yielded a triumphant wave of new information. Among these successes we must note the discovery of hosts of objects completely unknown until now, since they emit no visible light, or so little as to remain undetectable from below the atmosphere.

Furthermore, observations of objects that we *do* already know to exist, but made in wavelengths and frequencies never seen on Earth, have provided key insights into the objects' properties. Without going beyond the atmosphere, we would remain forever ignorant of whatever these objects can tell us by means of the infrared, ultraviolet, X-ray, and gamma-ray waves that they emit. As we shall see, Supernova 1987A provides a perfect example of the latter type of object: We would have seen it easily had we nothing but our eyes, but we know its true nature only because we now have gamma-ray, X-ray, ultraviolet, and infrared "eyes." Supernova 1987A offers a chance to see the workings of the "new astronomy," astronomical observations made above the Earth's atmosphere, once only a dream but now a reality.

3

THE OBSERVERS— THE EARLY DAYS OF SN 1987A

ON THE MORNING of February 24, 1987, Bob Kirshner arrived calmly at his office in the Harvard–Smithsonian Center for Astrophysics, an imposing set of interlocking buildings, built over an eighty-year interval on a small hill in Cambridge, Massachusetts. Perhaps one should not sneer at this hill for its modest rise of some sixty feet above the rest of the Harvard campus. The Harvard College Observatory, now part of Harvard's Center for Astrophysics, includes what was the world's largest telescope during the 1840s, a beautifully crafted 15-inch refractor that was sited on this hill in preference to the former observatory location, which now forms part of the Harvard Yard. Today the 15-inch telescope has only historical interest, and the not-so-lofty hill that it occupies draws its eminence from the distinguished collection of astronomers and astrophysicists who populate its buildings and constitute one of the most active and most productive groups of astronomical researchers.

Kirshner, a wiry, red-haired man just to the young side of forty, bears a vague resemblance to the entertainer David Letterman: His face shows the same animation even in repose, as if ready to leap into rapid verbal action once a suitable idea has surfaced (Figure 8). Kirshner is one of the most prominent observers of supernovae and their remants. On that February morning, his calm was based on temporary ignorance, which would soon be brusquely lifted.

Like many astronomers, Kirshner became fascinated by the

cosmos during his junior high school career, when he got his first good look at the heavens through a telescope. He attended Harvard College with the intention (possessed by about one student in every thousand) of becoming an astronomer, and went west to the California Institute of Technology in Pasadena for graduate work. Upon his arrival there in 1970, he encountered one of the best-liked Caltech astronomers, J. Beverly Oke (known to all as Bev), who asked him—as he asked all incoming graduate students—"So what do you want to study in astronomy?" Kirshner, like many new students, hesitated over the embarrassment of riches that lay before him. As a college junior, he had done a term project on a supernova remnant called the Crab Nebula. "How about supernovae?" Kirshner asked. Bev Oke said that was a good idea, and Kirshner proceeded to develop a graduate program leading toward such a specialty. He was installed in an office considered suitable for graduate students, and there he

FIGURE 8. Robert Kirshner of the Center for Astrophysics in Cambridge, Massachusetts, is one of the world's experts at observations of supernovae. (Harvard University Office of Public Affairs)

found as one of his neighbors the gray eminence of Caltech astronomers, the Bulgarian-born, Swiss-trained Fritz Zwicky, who had been at Caltech since 1925.

THE MAD SWISS

Zwicky, then in his early seventies, was a figure of some controversy and rancor. He had no hesitation whatsoever in proposing theories that seemed completely untenable, nor in asking anyone he chose questions that seemed incredibly foolish or contrived, nor in assuring his colleagues that their ideas were, at best, foolish and misguided. On closer examination, many of Zwicky's theories and questions remained untenable, foolish, and contrived. Some proved to have merit, but only years later, and since few astronomers had the patience to put up with Zwicky's caustic behavior (let alone his strange accent), he was often referred to (not in his hearing) as "the mad Swiss." Zwicky slowly acquired a status among more conservative faculty members as the Caltech house madman, a sort of astronomical Quasimodo, best kept behind several closed doors. Like Kirshner, Zwicky had an office in the second sub-basement of the Caltech astronomy department, where the room numbers all began with a double zero, as if James Bond had a hand in secret projects unrevealed to the world at large. There, in the quiet, Zwicky and Kirshner were free to interact as they chose.

Kirshner, although in some ways a typical young American scientist, had one habit that won Zwicky's heart: He arrived at work before anyone else. The reason for this was simple. Kirshner had married his high school sweetheart, who had become a substitute teacher in the Pasadena schools and therefore received early morning telephone calls informing her of her daily assignment. Once awakened by such calls, Kirshner found it expedient to get up and going. Zwicky gave Kirshner a piece of sound advice: "Always get to work before the Americans do." He also posed Kirshner scientific conundrums as the mood took him, which Kirshner took the trouble to ponder, occasionally to discuss, sometimes to solve.

Kirshner also learned that Zwicky was one of the world's experts on supernovae; in fact, were such names used in astronomical circles, he might have been known as Mr. Supernova. It was Zwicky's observations that had led to the classification of supernovae as either Type I or Type II. Not content with a mere two types, Zwicky had proceeded to categorize supernovae as Types III, IV, V, and beyond; some of these classes had only a single representative, and the ones beyond Type II never gained acceptance as entire classes. Instead, astronomers consider them unclassifiable, supernova oddballs.

Not merely an observer of the heavens, Zwicky had also engaged in investigations of the mechanism that could produce such titanic events as supernova explosions. In 1934, he had collaborated with Walter Baade, an astronomer at the Mount Wilson Observatory, likewise of European origin, to write two scientific papers on supernovae. Baade and Zwicky in fact originated both the name and the concept of supernovae. Their two papers lay almost forgotten for decades—until later advances in theory and observation showed how prescient Zwicky and Baade had been.

The first of the Baade–Zwicky papers analyzed the data on supernova explosions, and showed clearly that supernovae must be entirely distinct from the lesser events called "novae." Novae are stars that flare up, either once or several times, but in relatively modest outbursts that leave the star intact. Baade and Zwicky estimated the total light output from a supernova and showed that a supernova releases far more energy than any mere nova does. The astronomers concluded their first paper with the statement that "the phenomenon of a super-nova represents the transition of an ordinary star into a body of considerably smaller mass." They had glimpsed the essence of what causes most supernovae: the collapse of a star's core that initiates the explosion of the star's outer layers.

In their second paper, Baade and Zwicky analyzed the data concerning the mysterious "cosmic rays," which we now know to be relatively ordinary particles, such as electrons and protons, that have somehow been accelerated to speeds close to the speed of light. Baade and Zwicky concluded that "cosmic rays are produced in the super-nova process," for nothing else could produce so many particles moving at enormous velocities. Finally, at the end of this second paper, the authors stated that "with all reserve we advance the view that a super-nova represents the transition of an ordinary star into a *neutron star,* consisting mainly of neutrons. Such a star may possess a very small radius and an extremely high density."

This view, far from being reserved, represented an amazing leap of penetrating thought, for the neutron itself had been discovered less than two years before. By mind power alone, Baade and Zwicky had reached a correct conclusion about some, perhaps most, supernovae: Their cores turn into the amazingly compact objects that we now call "neutron stars," objects that are roughly the size of San Francisco, but made almost entirely of neutrons. In 1934, no one took neutron stars seriously as even potentially real objects. Even thirty-five years later, when "pulsars" were discovered and hypothesized to arise from rotating neutron stars, many astronomers still thought it unlikely that neutron stars could exist in the numbers suggested by Zwicky. But on this matter, it seems that Zwicky was entirely correct.

Equally prescient was Zwicky's work on the gravitational bending of rays of light, a phenomenon predicted by Albert Einstein in 1916, when he constructed the general theory of relativity to describe the behavior of matter in the presence of a gravitational field of force. Einstein had thought primarily of the sun's bending of light rays that pass close by it—an effect first observed in 1919, three years after Einstein's prediction, and which elevated Einstein from scientific prominence to worldwide fame. In 1937, Zwicky had studied this gravitational bending of light on a far larger scale: He asked what would happen if an entire *galaxy* of stars acted on the light from a still more distant galaxy. Zwicky correctly predicted that the entire galaxy would act as a "gravitational lens," focusing the light from the faraway galaxy, so that observers on Earth would see much brighter points of light than would otherwise be the case. The discovery of these gravitational lenses during the 1970s and 1980s has provided a full (posthumous) confirmation of Zwicky's calculations.

THE PATH TO SUPERNOVA 1987A

During the late 1960s and early 1970s, working two stories below ground, Bob Kirshner absorbed Zwicky's strange statements, along with the more conventional teachings of the other world-renowned Caltech professors. He wrote a Ph.D. thesis on his observations of supernovae and supernova remnants (wisps of gas and dust blown from exploding stars). In 1974, the year of Zwicky's death, Kirshner became a junior astronomer at the Kitt Peak National Observatory in Arizona, where he met Roger Chevalier, an astronomer more interested in the theoretical side of how supernovae actually explode.

In the mid-1970s, Kirshner and Chevalier embarked on a plan of observation to learn more about the details of supernova explosions. Since one cannot plan ahead to observe an explosion, the two astronomers concentrated on supernova remnants, which had the virtue of being known to exist. Kirshner's reputation as a supernova expert proved sufficient to gain him a faculty position at the University of Michigan, where he taught from 1976 until 1985, working not only on supernovae but also on observations of galaxies and galaxy clusters. Then, in 1985, Harvard called Bob Kirshner home (as Harvard would see it) to join its faculty. There Kirshner has continued his supernova observations and theoretical research while fulfilling his teaching duties (which have included, I am glad to say, the use of one of the textbooks I have written—a tribute relatively few astronomers have paid, but for which, as a fellow Harvard graduate, I am all the more appreciative).

On the last Tuesday morning of February 1987, Kirshner received a telephone call from Craig Wheeler, a theoretical astrophysicist at the University of Texas whom he knew well. In the 1970s and 1980s, Wheeler had performed important calculations (among many others) on the neutron stars that some supernovae leave behind—Zwicky's ideas brought up-to-date. "Have you heard about the supernova in the Large Magellanic Cloud?" Wheeler asked.

Kirshner was dubious. Several years earlier, he had been the victim of an astronomical practical joke along similar lines: While Kirshner was attending a conference in Erice, a marvelous site in western Sicily, close to an ancient ruined temple to Eros, some slightly crazed colleagues had sent him a telegram reading "Supernova in M51 [one of the closest galaxies to our own]; return home immediately." Kirshner had started to pack until one of those in on the joke told him the straight story. Come to think of it, Wheeler had been at that meeting. So Kirshner expressed considerable skepticism, but Wheeler assured him "No, no, it's real—I heard it from McCall." (McCall was a former student of Wheeler's, now returned to the University of Toronto, where the news from Chile had arrived.)

Kirshner decided that this merited action. He walked through the maze of corridors linking the old and new buildings at the Center for Astrophysics, past the granite pier of the 15-inch telescope, and reached the office of Brian Marsden, a British astronomer, expert on orbits in the solar system, but also known for his supervision of the bureau of astronomical telegrams—in effect, the worldwide clearinghouse for astronomical news. Something was afoot: Marsden was simultaneously talking on the telephone and typing on his computer keyboard. When Kirshner got his attention, Marsden informed him that indeed a supernova had been seen the night before, in the closest galaxy to the Milky Way.

Having received the shortest telegram that Marsden sent that day, Kirshner returned elated to his office. He decided that the supernova called for observations with his favorite astronomical instrument, the "International Ultraviolet Explorer" or "IUE" satellite. Kirshner had just decided to call the scientists in charge of the IUE satellite when his telephone rang: They were calling him. "What should we do about the supernova?" they asked. "We'd better get on it," Kirshner replied. He realized that the supernova was well placed in the sky for easy viewing by the IUE (which cannot observe all areas of the sky with equal ability), and arranged to work out an observing program for the next weeks.

ULTRAVIOLET OBSERVATIONS OF SUPERNOVA 1987A

Why was Kirshner so fond of the International Ultraviolet Explorer satellite for observing supernovae? In the days immediately following their initial outburst, supernova explosions produce copious amounts of ultraviolet waves. We may recall that these waves are akin to light waves, but shorter in wavelength and invisible to the human eye. As an exploding star's outer layers expand into space, the hot gas in them cools and gradually ceases to produce ultraviolet radiation. However, for a few weeks after a star explodes, if you want to study supernovae, you ought to study their ultraviolet as well as their visible-light output.

But there's a problem: Ultraviolet waves cannot pass through our atmosphere. In order to observe the universe in ultraviolet, you must pass beyond our atmospheric veil: You need a satellite. Kirshner had already used the IUE satellite, with its 18-inch reflecting telescope and its specialized ultraviolet detectors, to observe ultraviolet radiation from supernovae in far-distant galaxies. These observations suffered from the fact that because the technology used to produce detectors of ultraviolet has not advanced so far as the technology used for visible-light detectors, ultraviolet observations with the IUE are relatively difficult. Hence the supernovae in distant galaxies could not be seen in as much ultraviolet detail as astronomers would like; they continued to hope for a supernova much closer to Earth, which could be studied far more closely simply because its lesser distance would make it easier to detect.

Supernova 1987A offered just such a chance. The IUE could observe it easily, since the satellite circles the Earth in a "synchronous orbit," always remaining directly overhead of a particular point on the equator and capable of observing the entire sky. This satellite, launched in 1978 and still performing brilliantly, can observe cosmic objects in their visible-light emission as well as in their ultraviolet radiation, and can therefore compare the two directly. Without such a satellite, astronomers would be "blind" in the ultraviolet; with it, they can study the cosmos in ultraviolet radiation, not so well as they can in visible-light radiation, but well enough to satisfy them for now.

The IUE is directed by a team of scientists and technicians at NASA's Goddard Space Flight Center in Greenbelt, Maryland. An astronomer such as Bob Kirshner "observes" with the IUE by submitting a proposal that outlines what he hopes to see, and why such observations should prove useful. (Note that this makes proposals to look for what we *don't* know about unlikely to gain approval.) If the proposal passes muster, the satellite is directed to perform the requested observations, and the data, sent by telemetry to satellite headquarters in Greenbelt, are passed to the observer for analysis and distribution.

Over the years, Kirshner had proven successful at obtaining approval for his proposals. Kirshner owes his success to a combination of factors. These factors included his well-established credentials (Harvard, Caltech, the Kitt Peak National Observatory), his "track record" of publication, which showed that he did not "sit on his data," but instead analyzed and published it with reasonable speed for the rest of the astronomical world to examine, and his willingness to master the bureaucratic maze of proposal approval, making sure that at each step of the evaluation process his proposals were not sidetracked by inertia, jealousy, or plain old well-meaning inactivity. Using these and other assets, Kirshner had become one of the world's experts on supernovae, a Zwicky without the rough spots or sudden flashes of brilliant insight.

By the end of the day on February 24, Bob Kirshner knew that he had a once-in-a-lifetime astronomical opportunity before him, one that would last for several months, while the visible and ultraviolet radiation from the supernova slowly faded into invisibility. During this time, Kirshner and others could hope to learn what elements existed in the exploding star's outer layers, and how rapidly they had been blasted into space; furthermore, from changes in the data, the astronomers could hope to determine the evolution of an exploding star in detail never before obtainable. In the days and weeks following February 24, Kirshner became a prime operator in coordinating IUE observations of the supernova, and also joined in the quest for the answer to an obvious question: Which star had exploded? Did this star correspond to the candidates for supernova explosions considered likely by theoreticians?

SEARCHING FOR THE STAR THAT EXPLODED

Even more than most scientists, astronomers know the virtue of keeping records: You never know when you'll want to find out how things looked in the past. Once astronomical photography became commonplace, nearly a century ago, astronomers began to store their photographic plates, partly because each plate contains more information—images of thousand of stars and galaxies—than any astronomer can examine, or wants to, on a specific project. The largest plate collection, at the oldest university in the United States—Harvard—contains close to one hundred thousand separate images of the sky. These plates represent the accumulation of eighty years' work by several hundred astronomers. Some areas of the sky have been photographed repeatedly, others only a few times.

During the early 1950s, the National Geographic Society helped to sponsor a complete survey of all the sky visible from the Palomar Mountain Observatory, which was then the world's premier astronomical observing station. A team of astronomers led by George Abell spent two and a half years photographing the night skies with the newly installed 48-inch Palomar Schmidt telescope, a wide-angle instrument that can photograph a field of view 6 degrees across (twelve times the apparent diameter of the full moon), so that "only" 800 photographs are needed to cover the entire sky as seen from Palomar Mountain. (The 200-inch Palomar reflector has a field of view that spans only one one-hundredth of the area photographed by the 48-inch, so it would require 80,000 photographs to cover the entire sky—an impossibility in a human lifetime, and a rather poor use of the giant telescope.)

The "Palomar Sky Survey" that emerged from Abell's work included two photographs for each observed region of the sky, one taken with a filter that admitted only blue light, the other with a filter that let only red light onto the photographic plate. This tandem allowed astronomers to determine, in a rough sense, the colors of the objects on the photographs—which ones emit mostly red light, which mostly blue light, and which emit roughly equal amounts of these two colors of visible light. For three decades, the Palomar Sky Survey has been a mainstay of astronomical research. The discovery of any new object, or type of object—"quasars," "pulsars," "X-ray emitting binary stars"—immediately sends astronomers to the Sky Survey, asking questions such as, How did the object appear a few years ago? Has it changed its light output? How many other similar objects appear on a given plate of the Sky Survey?

But the Palomar Sky Survey has one flaw: It covers only two-thirds of the sky, and provides no view of the southernmost portion of the heavens, the part that never rises over southern California—the part of the sky where Supernova 1987A appeared. When astronomers learned that Supernova 1987A had exploded in the Large Magellanic Cloud, they knew at once that the Palomar survey was of no use to them, and that they would have to make do with other, less inclusive surveys of the southern skies.

As it happened, however, Nicholas Sanduleak, an astronomer at Case Western University in Cleveland, Ohio, had made a detailed survey of the blue stars in the Large Magellanic Cloud several years earlier. Looking at Sanduleak's list of stars, each one designated by "Sk" (for Sanduleak) and the star's astronomical coordinates, astronomers immediately saw that one star on the list, Sk −69° 202, had the same position on the sky as the supernova. But was 202 the star that had exploded?

The answer to this question lay within a tangled web for several weeks, because the IUE has only a modest ability to see detail on the sky, that is, to discriminate between nearby points of ultraviolet emission. Photographs of the Large Magellanic Cloud before the supernova outburst showed two stars that should have appeared in the IUE's field of view, 202 and one other. Since a supernova produces ultraviolet light for only a few weeks, astronomers expected that once the supernova had faded into ultraviolet invisibility, the IUE would observe only one star in this area of the sky. But even after the supernova's ultraviolet radiation had decreased below the threshold of detection, the IUE continued to observe two stars! Taken at face value, this result implied that 202 was still shining, and that a third, previously undetected star must have exploded.

The confusion was lifted with a triumph of what astronomers call "classical astronomy"—the old-fashioned examination of photographic plates, especially those from the 4-meter telescope at the Cerro Tololo International Observatory in Chile, which were analyzed by Barry Lasker and Nolan Walborn. Using these plates, "classical" astronomers could determine that not two but *three* stars had existed within the IUE field of view, and that the "missing" star—the one that had exploded—was indeed number 202. Thus once the supernova had faded from view, the two stars still observed by the IUE were the two faint neighbors of old 202. The third star that provided the key to the puzzle was quite faint—so faint that it could barely be seen on the photographs—but nevertheless produced sufficient ultraviolet radiation to be the second star that the IUE saw after the supernova had faded. Because the old photographic plates made it clear that three stars had existed before the explosion in the field of view, and because two stars plus the supernova existed there after February 23 (one

ic charge, equal in amount but opposite in sign to the
ive charge on a proton (see Figure 9).

addition to the protons, neutrons, and electrons that form
ns, the universe contains other types of particles. *Photons,*

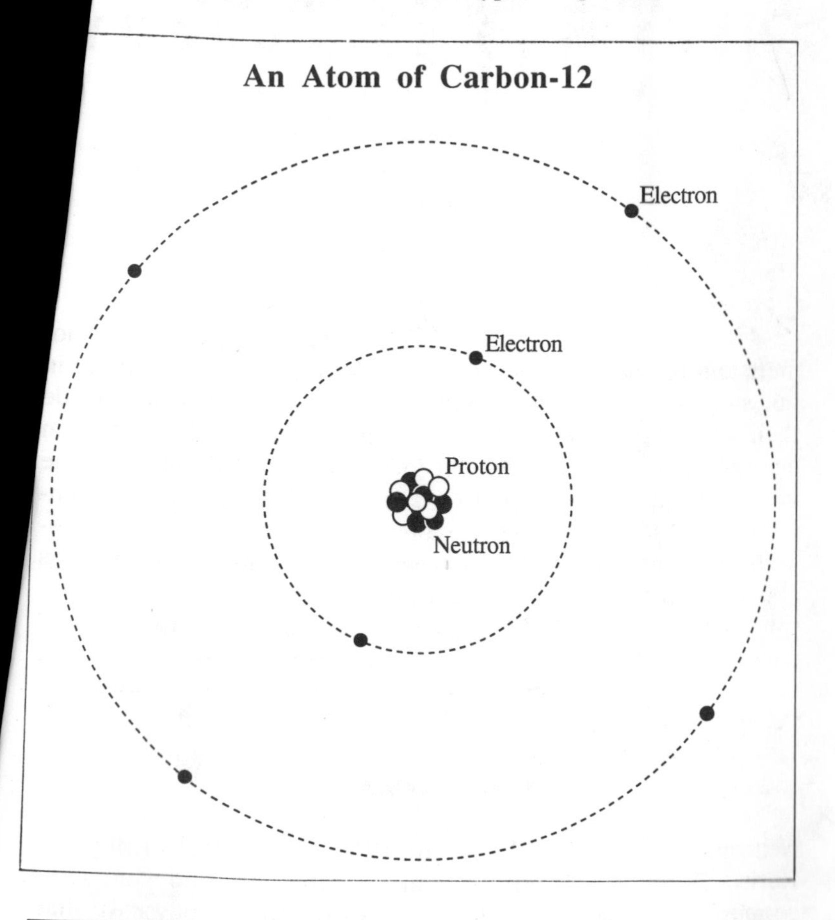

FIGURE 9. An atom of carbon-12 consists of a nucleus of six protons and six neutrons, around which orbit six electrons. Two of the electrons orbit in an "inner shell," and four move in a larger orbit. If this were a scale model, the protons and neutrons would be the size of ball bearings, and the electrons would be microscopic particles orbiting at distances of several miles from the nucleus. (Drawing by Marjorie Baird Garlin)

of the stars too faint to be detected with the IUE, but present all the same), by March 1987, the experts had concluded that the supernova had indeed exploded from the previously observed star −69° 202.

THE MYSTERY SPOT

So the identification appeared complete: The blue star 202 had indeed exploded. But a further mystery arose when, during the spring of 1987, three astronomers from the Harvard–Smithsonian Center for Astrophysics, Peter Nisenson, Kostos Papaliolios, and Margarita Karovska, went to the Cerro Tololo Observatory in Chile to observe the supernova, using an advanced technique called "speckle interferometry."

Speckle interferometry attempts to correct for the blurring of any astronomical image that the Earth's atmosphere produces. Each small parcel of air acts as a lens, deflecting light rays slightly from what would otherwise be straight paths through the atmosphere. Because of air currents in our atmosphere, the direction and amount of this deflection continuously changes. As a result, every astronomical object appears to us not as a point but as an image spread over a small area of the sky. Speckle interferometry allows astronomers to record the changing distortion caused by the atmosphere on an extremely rapid time scale and to correct the image to remove some of the effects of this distortion. Thus astronomers using speckle interferometry can record more detail than any single, momentarily changing image can provide.

The astronomers from the Center for Astrophysics applied this advanced technique to the supernova. On May 4, 1987, they announced that they had found another image quite close to the image of the supernova. This second image was dubbed the "mystery spot" by Bob Kirshner in honor of the tourist attraction of the same name that he'd seen during a visit to Santa Cruz, California, the previous summer. The mystery spot produced about 10 percent of the supernova's brightness. If the image indeed represented an object at the supernova's distance, it must be about 260 billion miles from the supernova—the distance that light travels in seventeen days. What could such a bright object—because even one-tenth of a supernova's brightness is immensely bright—be doing there?

The answer to date is simply that no one knows. Some astronomers take a conservative approach, and believe that the mystery spot represents some artifact of the imaging process, not a real object. Other astronomers speculate that the supernova explosion encountered a cloud of gas and dust, perhaps material ejected from star 202 before it exploded. This explanation would require that the shock wave travel at one-half of the speed of light, since the mystery spot was found only about thirty days after the supernova was detected, implying that the explosion reached this hypothesized cloud, seventeen light-days from the star, thirty days after the initial outburst. (To travel in thirty days the distance that light travels in seventeen days is to travel at about one-half the speed of light.)

Some theoreticians have suggested that the exploding star ejected a stream of material traveling at nearly the speed of light. This jet would have struck nearby gas and dust, heating it until it glowed with the light that produced the mystery spot. Since the mystery spot has now long faded from view, we may never know which theory has the best chance of proving correct.

Far more exciting, however, than any mystery spot close to SN 1987A are streams of particles that we *do* know the supernova emitted. The most fascinating puzzle to emerge from our observations of the supernova deals with the tangled web of observations of a particle never before seen from an exploding star—what physicists call the wily neutrino.

4

THE GROWTH NEUTRINO ASTR

TO A SMALL cor mers and physicists, Supernova 1987A marked a tu our study of the universe. With SN 1987A, human their first observation of an object beyond the sol *neutrinos,* tiny particles thought to be incredibly abun universe, but so elusive as to evade detection with a plete success. Neutrinos have come to be recognized icists) as one of the most basic types of particle in the u abundant as the photons that form electromagnetic w spite this fact, and despite the tremendous (hypothesize dance of neutrinos in the universe, we are still at the stag every detection of a neutrino from a particular cosmic ob mains a rare event.

WHAT IS A NEUTRINO?

Neutrinos were named during the 1930s by the Italian phy Enrico Fermi, a leading figure in the effort to understand cosmos at the smallest levels of size. By then, physicists come to believe (as they still do) that all familiar types of mat consist of atoms in which one or more particles called *electro* orbit around a central nucleus. The nucleus consists of two typ of particles: *protons,* each of which carries a positive electri charge, and *neutrons,* each of which carries no electric charge The electrons in orbit around the nucleus each carry a negative

the carriers of electromagnetic waves, are one such type. We know that photons exist because we can easily detect them, and by the 1930s physicists knew that a photon has no electric charge. Thus, like a neutron, a photon is electrically "neutral" (uncharged). In Italian, the neutron has the name "neutrone," or "big neutral thing." Thus, when the need for a name arose to describe yet another particle with no electric charge, Fermi coined the name "neutrino" ("little neutral thing"). Fermi's name gained worldwide popularity, so today when we describe the mysterious, evanescent neutrino, we employ an Italian word.

But where do we find these italonymous particles, the neutrinos? Neutrinos have the distinction of being *invented* before they were *discovered.* By the early 1930s, the Austrian physicist Wolfgang Pauli had perceived that in order to explain the results observed when certain types of atomic nuclei "decay" to form other types, an *undetected* particle must exist! This particle left no track or trace in the instruments that detected familiar types of particles, such as nuclei and electrons. Nevertheless, Pauli saw, because the nuclei lost energy when they decayed, this invisible particle must have been created as part of the decay process. Thus Pauli created in his mind a particle that could not be detected directly.

THE DETECTION OF NEUTRINOS

An essential aspect of the nature of neutrinos is that they barely interact with familiar forms of matter, such as atoms and their nuclei. Hence the actual detection of neutrinos occurred more than twenty years after Pauli first "saw" them, and almost as long after Fermi had named them. In 1938, immediately after receiving the Nobel Prize, Fermi left Italy for the United States. (Influenced by his experiences in Mussolini's Italy, Fermi turned part of his Nobel Prize money into gold coins, and kept them hidden under the floorboards of his new American home.) During the Second World War, Fermi played a key role in developing the atomic bomb by supervising the first controlled chain reaction of uranium nuclei, which took place on December 2, 1942, beneath the grandstand of Stagg Field at the University of Chicago. Sadly for science, Fermi died at the age of fifty-two, in 1953, still an immensely productive physicist. A few years later, and with great difficulty, physicists finally managed to detect neutrinos produced in particle accelerators. Even today our most sensitive equipment can detect only one in trillions upon trillions of the neutrinos that pass each second through the Earth—and everything else.

Most of the neutrinos from space come from the sun, which emits great numbers of neutrinos every second in its inner core, where nuclear fusion occurs. Almost all of these neutrinos, traveling at the speed of light, pass through nearly half a million miles of solar material without the slightest interaction. Those that happen to be directed toward the Earth likewise pass through the solid planet on which we live with nearly no consequence. Billions of neutrinos pass every second through each of *us,* luckily without effect.

But suppose that you build a neutrino detector, and place it far underground (in order to screen out the effects of other particles from space that can't pass through matter as easily as neutrinos). Then you can detect several neutrinos per hour from the sun, and you may—with luck—detect a few of the neutrinos that traverse the cosmos. Many of these are thought to arise in supernova explosions. Supernovae should be great sources of neutrinos, because calculations of how stars work imply that great floods of neutrinos must arise as a star's core undergoes the collapse that initiates the supernova explosion. But could this prediction of a neutrino flood, which forms a keystone of every theoretical model of supernova explosions, ever be directly demonstrated? Could the burst of neutrinos from a star's collapsing core be detected?

NEUTRINOS FROM STELLAR HELL

The answer is yes. Two weeks after the first news of Supernova 1987A, Bob Kirshner sat in his office at the Harvard–Smithsonian Center for Astrophysics, pondering how best to take advantage of his good fortune. His telephone rang: It was John Bahcall, a respected theoretician from the Institute for Advanced Study at Princeton, in town to give a colloquium at the Massachusetts Institute of Technology. As Kirshner tells it, Bahcall asked him, "How do I get from Harvard Square to the Observatory?" This struck Kirshner as an odd question, since Bahcall had been a student at

Harvard, but Kirshner realized that he was dealing with a theoretician, and answered it.

Bahcall arrived at Kirshner's office and borrowed Kirshner's razor in order to have a shave before presenting his colloquium. This presentation was to deal with the elusive neutrinos, the particles that lie at the hub of Bahcall's scientific career. Bahcall ranks among the world's experts on the mechanisms by which the sun and other stars produce neutrinos at their centers, and on the ways that such particles might be detected despite their almost complete reluctance to interact with other forms of matter.

John Bahcall is an introspective, hardworking astrophysicist, a man who grew up in the small Jewish community of Shreveport, Louisiana, entered the University of California at Berkeley with the intention of becoming a rabbi, was converted to science during his junior year, did his graduate work at the University of Chicago and at Harvard, and was a junior professor at Caltech before joining the Institute for Advanced Study. By now, having reached his early fifties, Bahcall's face shows the creases of an intense, gregarious man who has derived great pleasure from scientific thought.

On March 10, 1987 Bahcall was emerging from two of the most intense weeks in his life, two weeks filled with calculations, hopes, rumors, and finally confirmed reports. The calculations dealt with the number of neutrinos produced by a supernova, of the number of neutrinos that might reach the Earth, of the efficiency of the Earth's neutrino detectors. Bahcall had heard about the supernova on February 24, a few hours after Brian Marsden had given Kirshner the news. Most of the scientists who work in universities or university-affiliated research institutions routinely send messages to one another via "Bitnet," a national electronic computer linkage system that allows each of the thousands of users to send messages ("electronic mail") to whomever they choose. (The name refers to the "bits" of information that computers process, but arose, with typical computer humor, as the acronym for "because it's there.") For two weeks, Bitnet had become Bahcall's line of consciousness, as he sent and received message after message to and from those in the "neutrino game," using the telephone only for his most complex interactions with his fellow scientists. Dozens of experts on exploding stars, on the neutrinos they might produce, and on the means of detecting and analyzing such neutrinos, had joined Bahcall in recognizing the rare opportunity that confronted them. Some of them, most notably Adam Burrows of the University of Arizona, had spent years in calculations of just how many neutrinos should emerge from a supernova. Now their opportunity lay before them—or more precisely, had just passed.

For John Bahcall, the explosion of Supernova 1987A raised the following pressing questions: Had the supernova produced enough neutrinos to make the Earth's neutrino detectors respond? And if so, had the detectors in fact registered neutrinos from the supernova, or had the detectors—from malfunction or any other reason—missed the neutrino moment of a lifetime?

To answer the last question, you need a good neutrino detector. And by tremendous good fortune, not one but *four* giant neutrino detectors were in operation on the day that the supernova was detected. Intriguingly, the two best "neutrino detectors" were not designed for neutrino detection. They had another goal—to test a popular theory of modern physics. The theory predicts that every once in a while, one of the protons at the heart of an atom should spontaneously "decay," turning itself into other sorts of particles. (This "once in a while" is about once every billion trillion trillion (10^{33}) years for any individual proton, so you need lots of protons to observe this effect on shorter times scales.) Therefore the "neutrino detectors" were searching not for neutrinos from space but for the products of "proton decay" in their immediate vicinity. But because these instruments can also detect neutrinos, they provided the world's most sensitive equipment for finding neutrinos from cosmic sources. On the morning of February 23, 1987, they detected neutrinos arriving from a distance of 160,000 light-years.

THE DETECTION OF NEUTRINOS FROM SUPERNOVA 1987A

All four of the neutrino detectors lie deep underground, placed there in order to avoid confusing the detector with other types of particles, called "cosmic rays," that bombard the Earth's surface from outer space. The cosmic-ray particles cannot penetrate half a mile of solid Earth, so the detectors were constructed in old salt mines, or in tunnels that pass beneath entire mountains of rock, to shield them from the particles reaching the Earth's exterior. But no amount of rock can shield you against neutrinos: Because neutrinos interact so rarely with "ordinary" matter, they pass through the entire Earth as if it barely existed, and it takes a huge amount of matter in your "neutrino detector" to stop even one neutrino in a trillion trillion. Indeed, the neutrinos observed from the supernova were *leaving* the Earth: They must have passed *upward* through the Earth, not downward from the visible sky, since the supernova never rose above the horizon at the latitude where the four detectors operate.

One of the four detectors, the IMB (whose abbreviation honors the University of California at Irvine; the University of Michigan; and the Brookhaven National Laboratory on Long Island), occupies a cavern in the Morton Salt mine, half a mile underground near the edge of Lake Erie in Ohio. A second detector, called Kamiokande, was constructed deep underground in Kamioka, Japan (the "nde" at the end stands for "nucleon decay experiment"; a "nucleon" is a proton or neutron). A third neutrino detector, the "Mont Blanc detector," is an Italian–Soviet collaborative effort, located in a chamber near the middle of the Mont Blanc auto tunnel, underneath a mile or two of rock that forms part of the western Alps. The fourth neutrino detector, the "Baksan detector," lies inside a tunnel in the Baksan river valley of the Caucasus mountains in the Soviet Union.

Both the IMB and Kamioka detectors contain many tons of water—two gigantic swimming pools buried deep underground. Suspended at intervals throughout the pool, and along the walls that bound it as well, are photomultiplier tubes, which can detect flashes of light called "Cerenkov radiation." Cerenkov radiation arises only when an electrically charged particle travels through some medium (in this case water) at a speed *greater* than the speed of light in that medium. You can never produce Cerenkov radiation in an empty space, since light travels through empty space at 186,000 miles per second, the greatest speed attainable in the universe, and thus more rapidly than any particle can. But in a transparent medium such as water, light typically travels more slowly than it does through a vacuum. In this case, a particle moving at nearly the speed of light can have a speed *greater* than the speed of light in that medium! The particle then acts like the prow of a speedboat, and the Cerenkov radiation resembles the bow wave that ripples through the water, produced by the boat's fast-moving prow.

In order not to lose the faint glow of Cerenkov radiation in the murk of a swimming pool, the IMB and Kamioka detectors contain the world's purest water. At the IMB detector, this water resides within a quarter-inch-thick sheet of black polyethelene plastic, an enormous water bed for science (Figure 10). This water has been so effectively cleansed of all pollutants that the entire tank, a cube of water nearly seventy feet on a side, remains crystal clear. The scuba divers who service the pho-

FIGURE 10. The Irvine–Michigan–Brookhaven (IMB) detector contains 7,000 tons of highly purified water, kept half a mile underground in a tank lined with 2,000 light sensors. These sensors can detect the Cerenkov radiation produced when a neutrino interacts with a particle in the tank of water, an event that occurs only rarely. (Lawrence Sulak, Boston University)

toelectric detectors suspended within the IMB tank at three-foot intervals enjoy greater visibility than skin divers on the Barrier Reef. Any particle moving faster than the speed of light in water that passes through the 7,000 tons of water will create a flash of Cerenkov radiation. The detectors record these flashes with nearly 100-percent efficiency. For some of the flashes, the physicists can reconstruct the path of the particle that zipped through the water at nearly 186,000 miles per second.

The Kamioka detector was originally constructed to detect high-energy neutrinos that might arise from proton decays, and had been modified to be capable of detecting neutrinos from supernovae only a few months before the neutrinos from SN1987A completed their 160,000-year-long journey to Earth.

Unlike the IMB and Kamioka detectors, the Mont Blanc detector under the Alps and the Baksan detector beneath the Caucasus mountains use a special mixture of gallium and chlorine, not liquid water, to detect neutrinos. However, as Bahcall's calculations showed, all four detectors should—in theory—have been able to detect some of the neutrinos from the supernova, even though those neutrinos had spread out into an enormous volume of space as they traveled 160,000 light-years of distance from the Large Magellanic Cloud. Because the blast wave that produces the *light* from the supernova is temporarily trapped within the exploding star's outer layers (which the neutrinos penetrate with scarcely an interaction), the neutrinos can escape from the supernova several hours before the initial burst of light emerges from the star's surface. Therefore, since neutrinos travel through space at the speed of light, the neutrinos should have reached the Earth and been available for detection a few hours before McNaught's photograph recorded the supernova in Australia on February 23.

When John Bahcall received the news of SN 1987A, he did everything within his power to encourage his fellow scientists who operated the neutrino detectors to analyze their data as quickly as possible. The neutrino detectors do not ring a bell each time a neutrino passes through, for despite the shielding provided by half a mile of earth, they are continually bombarded by various types of particles, not just by neutrinos. Extensive data analysis must be made in order to determine which particles struck at a given time. Typically, the detectors register light flashes (presumably Cerenkov radiation) automatically, storing their data on magnetic tape for later computer analysis, which may occur days or weeks later.

The Japanese scientists operating the Kamioka detector were the first to realize that they might have detected neutrinos from Supernova 1987A. They notified the IMB group, who searched their records and found an apparent burst of neutrinos within one minute of the time that the Kamioka detector had recorded its burst. Almost immediately thereafter, the Mont Blanc and Baksan detector teams looked through their records—and found an intriguing result.

Within a few days after the news of the supernova, Bahcall and his associates, drawing on previous work by other neutrino experts, had written a paper predicting how many neutrinos should have reached the Earth, and had shown that the four neutrino detectors had a reasonable chance of having recorded them. Reacting with unusual speed, the editors of *Nature,* one of the most respected scientific periodicals, had this paper in print by the following week. Bahcall circulated his prediction still more widely and rapidly by sending an "IAU telegram," a bulletin distributed around the world through the auspices of the International Astronomical Union. Such telegrams typically announce new *observational* discoveries, and the telegram announcing the discovery of SN 1987A ranks among the most famous of these. Bahcall's IAU telegram, however, announced a *theoretical prediction,* not the sort of thing one usually circulates so widely, but despite his innate conservatism in scientific matters, Bahcall sent his telegram to the world of astronomers, seeking to increase scientific interest in the neutrinos that he was sure must have reached the Earth on the morning of February 23.

In Japan, the computer printer that recorded the analysis of the Kamioka detector temporarily failed from overwork, as the scientists there printed their results far more rapidly than usual—reams upon reams of records from the photomultiplier tubes, covering the relevant hours when neutrinos might have reached the detector. And with what result? Had the neutrinos from Supernova 1987A been detected?

This question proved far from trivial to answer. All four detectors recorded *something* on February 23. It would be pleasant to report that they all saw a burst of neutrinos from Supernova

1987A, but this is not the case. Eventually, the truth emerged, though not without argument against the scientists seeking that truth.

WHICH DETECTOR SAW THE NEUTRINOS?

Study of the automatically recorded data from the four neutrino-detecting sites revealed that each of them had detected a quick, sharp pulse of neutrinos on February 23. The trouble was, not all the detectors had detected a pulse at the same time!

Two of the four—the IMB and Kamioka detectors—*did* register a "spike" at effectively the same moment. On the morning of February 23, eight neutrinos were detected at the IMB within an interval of five and a half seconds. Normally only a few would be detected per hour, and their appearance would be spread out over time. Each of these particles interacted with a proton or electron in the water, creating a flash of Cerenkov radiation that registered on the IMB's photoelectric sensors. At Kamioka, Japan, the same sort of detector, but with even better sensors, yielded much the same result. Kamioka detected eleven neutrinos within an interval of a minute or less. Because the Kamioka detector has less definitive timing than the IMB, the correspondence of the arrival of the neutrinos at the Kamioka and IMB detectors cannot be made more precise than this "minute or less." However, the key minute of time at Kamioka did include the five-and-a-half seconds during which eight neutrinos passed through the IMB in Ohio. (In case you are wondering whether there should be a time difference between the arrival time of neutrinos in Japan and Ohio, simply because these two points must have had different distances from the supernova, recall that neutrinos travel at the speed of light, and it takes light one-twentieth of a second to travel from Ohio to Japan, or vice versa.) Thus the IMB and Kamioka results confirm each other.

Because the Mont Blanc and Baksan detectors have smaller volumes in which to trap neutrinos than the IMB and Kamioka detectors do, the neutrinos from SN 1987A lay just at their thresholds of detection: Each of them should have detected only one, or possibly two, neutrinos from the supernova. However, the Baksan detector recorded *five* events on February 23, all of them at nearly the same time as the two large, water-filled neutrino detectors. This surprisingly large number of events is a bit unsettling, but the coincidence in time is reassuring, and most scientists conclude that the Baksan detector probably did record neutrinos from SN 1987A.

But the Mont Blanc detector results pose a serious problem. The Mont Blanc experiment showed that five events had occurred in the detector within a single minute, but not at 2:35 A.M. Eastern Standard Time on the morning of February 23, the time when the IMB and Kamioka detectors recorded their neutrinos. Instead, at the Mont Blanc detector, the events were registered four and a half hours earlier, at 9:52 P.M. (EST) on February 22.

The theory of supernova explosions calls for the collapse of a star's core to produce a single blast of neutrinos. Hence, to the extent that we rely on the theory—and most astrophysicists believe quite strongly in its basic outlines—we expect a single pulse of neutrinos, all produced within a few seconds of each other. Nothing in space would slow these neutrinos as they spread in all directions through space, taking 160,000 years after the collapse that began the explosion of Supernova 1987A to arrive at Earth. We therefore expect all the neutrinos to reach the Earth at nearly the same time, not at intervals of several hours.

THE VERDICT: JAPAN AND THE U.S. YES, RUSSIA MAYBE, ITALY NO

Statistical analysis of all four detections leaves most scientists believing that *something* happened to produce the events that the Mont Blanc experiment detected, but that it was not SN 1987A. The mystery remains as to why the Mont Blanc experiment happened to detect a greater number of neutrinos than usual on the same day that the supernova was detected. The scientists involved in neutrino detection are prepared to accept this as not nearly so great a mystery as the notion that the supernova could have emitted two separate blasts of neutrinos, more than four hours apart. Hence, they conclude—at least outside Mont Blanc—that the supernova's neutrinos were recorded by the detectors in Ohio and Japan, and probably in Russia, but not in the Italian Alps.

COMING OF AGE WITH NEUTRINO ASTRONOMY

Neutrinos had been recognized (at least theoretically) as one of the most abundant and most significant types of particles in the universe, as important as the photons that form light and all other types of electromagnetic radiation. But as the year 1987 began, astronomers had yet to observe any cosmic object save our sun by the neutrinos it produced. Once the eight IMB neutrinos and the eleven Kamioka neutrinos that arrived within a minute were accepted as arising from the supernova, though, astrophysicists had something to crow about. For the first time, "neutrino astronomy" had detected an object other than our sun, which creates neutrinos as part of the nuclear fusion processes at its center. Indeed, because the neutrinos emerged from the supernova a few hours before the light waves did, if the underground neutrino detectors had been capable of real-time computer analysis and had been equipped with alarm systems, *they* could have "heard" the supernova explosion and announced it, since the neutrinos arrived many hours before Duhalde and Shelton saw the supernova in Chile.

Since we know the distance to the supernova, and we know the efficiency with which both IMB and Kamioka detect neutrinos, we can calculate the total number of neutrinos that the supernova produced in its few seconds of collapse. This number is so gigantic—one with fifty-eight zeros after it—that words fail to describe it, as is so often the case in astronomy. Perhaps as much as 99 percent of all the energy produced by the explosion emerged in this flood of neutrinos.

The Earth, at a distance of 160,000 light-years (about a billion billion (10^{18}) miles) from the supernova, received a blast of about ten thousand trillion trillion (10^{28}) neutrinos. Each person on Earth, presenting a surface area of about a thousand square centimeters toward the supernova (more if the person happened to have a direction perpendicular to the supernova's direction), had many trillion neutrinos pass through his or her body. Since the 4 or 5 billion people on Earth have a collective volume about 100,000 times the volume of the water in the neutrino detectors, and since our bodies stop neutrinos about as effectively as water does, each of several million people must have had a neutrino interact with his or her flesh on the morning of February 23,

1987. However, because neutrinos have no perceptible effect on human beings, these detections went unnoticed.

LIMITING THE MASS OF NEUTRINOS

On the afternoon of March 10, 1987, a clean-shaven, properly located John Bahcall gave his colloquium at MIT to a packed house. Never one to leap to conclusions, he spent most of the time discussing his specialty, neutrinos from the sun. Finally, he mentioned the apparent detection of neutrinos from the supernova, which had become certain only a few days before. Thinking about this detection, Bahcall had perceived an item of crucial importance, not original with him, but of great interest nevertheless: If the neutrinos indeed came from the supernova, their detection allowed scientists to place a significant upper limit on the *mass* of these neutrinos.

The mass of any particle measures the *amount* of matter that it contains. On Earth, we can measure mass by the force of the Earth's gravity on a particle, since a particle with more mass has a greater gravitational force acting upon it; the amount of that force yields what we call the particle's "weight." Scientists had long speculated that neutrinos have no mass at all, as is true of the photons that constitute light and other types of electromagnetic radiation, such as radio, infrared, and ultraviolet waves. Massless particles exist but weigh nothing. Nevertheless they permeate the universe in enormous numbers, and always travel at the speed of light, the maximum speed attainable.

Astronomers care a great deal about any possible mass of the neutrino, not so much because they love neutrinos but because they are deeply concerned about the *"missing mass"* in the universe. The phrase "missing mass" describes the mass that astronomers deduce to exist from the gravitational force that the mass produces, but which they have yet to observe directly. The amount of gravitational force is itself deduced from the motions of stars within galaxies and of galaxies within clusters of galaxies. Astronomers can measure how rapidly these objects are moving. Since gravity makes them move, and since gravity depends upon mass, they can deduce the amount of mass required to make them move at a given speed. When scientists perform such calculations, they typically find that ten to a hundred times more mass must exist than can be explained in terms of the objects that astronomers can *see*. Hence the "missing mass" must consist of "dark matter," which does not emit electromagnetic radiation such as light. In fact, the "dark matter" contains ten to a hundred times more mass than the mass that we observe as it shines in stars and galaxies. If it turned out that each neutrino had a tiny amount of mass, we could clear up the mystery. The "missing mass" would then become the "elusive mass" contained in neutrinos that fill the universe and bombard us by the billions without much interaction.

Neutrinos' remarkable unwillingness to interact with other forms of matter makes them extremely difficult to detect. For this reason, the upper limit on the mass per neutrino—the maximum mass established by experiment—has remained relatively high. This rather large maximum limit on the mass of neutrinos meant that we could not rule out an important possibility: that because neutrinos are so numerous in the universe, even a tiny mass-per-neutrino could still imply that the totality of neutrinos had a mass greater than the mass in any other type of particle. Thus, before SN 1987A, physicists recognized that neutrinos might indeed be the source of the "missing mass."

But a particle with mass, even a tiny mass, travels more slowly than a particle without mass, which always travels at the speed of light, close to 186,000 miles per second. If neutrinos have mass, we would expect that they would have taken longer to reach us than the light from Supernova 1987A. In fact, however, the neutrinos from the supernova arrived on Earth a few hours *before* the light—not because they traveled more rapidly than light, but because models of supernova explosions imply that they emerged from the supernova a few hours before the light. The neutrinos were produced during the initial *collapse* of the star's core rather than in the subsequent *explosion* that ripped through the star's outer layers. Calculations based on the principle that only particles with zero mass can travel at the speed of light now place an upper limit on the neutrino mass. This upper limit equals one part in 50 million of the mass of a proton, one of the types of particles in an atomic nucleus.

This result has an important impact on our notions of the entire universe. The new upper limit on the mass per neutrino effectively rules out neutrinos as the "missing mass." Of course, if neutrinos do have zero mass, they were doomed to fail as candidates from the start, but the observational confirmation nevertheless is significant.

The basic outline of this argument was familiar to Bahcall on the afternoon of March 10, and by putting numbers into his thoughts, Bahcall could already eliminate, to his satisfaction, the possibility that the type of neutrinos emitted in great numbers by a supernova could provide the universe's "missing mass." In one brief afternoon, the supernova had shattered hopes for that type of neutrinos to provide the explanation to questions that burn in the hearts of astronomers. What's the matter? Where's the "missing mass"? Where, in fact, is the bulk of the universe?

The detection of neutrinos from SN 1987A helped to tighten the existing limits on the possible mass of one type of neutrino, and thus to eliminate that type from the missing-mass particle sweepstakes. Two other types of neutrinos are known to exist, and their possible mass remains unaffected by the supernova results. In addition, a host of other particle types—types that are verified to exist and types that are completely speculative—remain "alive" as candidates for the missing mass. The missing-mass problem remains open, to puzzle, delight, and confuse humanity as we seek to understand the cosmos that surrounds us.

WHAT GOOD IS RESEARCH?

An intriguing sidelight on the detection of the neutrino blast from the supernova is this: The experiments that found the neutrinos never found any proton decays. In other words, machines that were designed to be proton-decay detectors turned out to be neutrino detectors only. But lest one conclude that the experiments failed in their basic mission, it is crucial to note that by *failing* to find any proton decays, the detectors *disproved* some of the theories constructed by particle physicists who aimed to improve and to simplify our understanding of the universe. Of course, theoreticians are not so easily daunted. They have proceeded to search for new and better, modified theories that will incorporate the negative results of the proton decay experiments, yet satisfy their longing—as noble as longings get in physics—to find the holy grail, a theory that will explain all types of forces in the universe as parts of a single force.

Meanwhile, the proton decay experiment accomplished the first observation of neutrinos from a supernova explosion—a perfect example of the kinds of unexpected side benefits obtainable from inquiries at the frontiers of science.

THE LONG CHAIN OF SUPERNOVA OBSERVATIONS

Within a few weeks of the supernova's detection, a well-organized observing program via satellite and ground-based observatories was underway. (In the case of the neutrino observations, the observations were over even before anyone knew it was time to begin them.) The data continued to pour in, affording astronomers who specialized in visible light, gamma rays, X rays, and infrared a chance to shine. As the news of the supernova spread first through the astronomical community and then the wider world beyond, one additional group of astronomers moved into high gear: the supernova theoreticians. This long-suffering bunch had never had a relatively nearby supernova to confirm or to refute their results. Now it was show time, a time to see whose reputation would rise like the initial spike of light from an exploding star, whose would fall like the slow decline that follows the initial outburst.

Before we tackle these problems, however, we ought to pay tribute to the supernovae that preceded SN 1987A, and to the astronomers who studied them. Although Supernova 1987A is the most important exploding star of this century, it has illustrious predecessors, some of them much closer to Earth than the Large Magellanic Cloud where 1987A resides. By studying these earlier supernovae we can learn a great deal, and can better understand the remaining mysteries that SN 1987A may help us to resolve.

5

SUPERNOVAE IN HISTORY

At rare intervals, for as long as humanity has been looking at the stars, supernovae must have burst upon human consciousness, disrupters of the seemingly eternal fabric of the heavens. Hence our history embraces an ever-growing knowledge of the universe in which we live. But we have paid a price for this knowledge. Long ago, we felt ourselves to be part of the cosmos, caught in its fabric, with the heavens draped close above us. Ironically, the truth revealed by astronomy—that enormous distances separate us from our cosmic neighbors—has made the cosmos seem less important to the lay observer. Though understandable, this attitude is foolish: We remain part of the universe no matter what distances separate and connect us. And as our studies of supernovae have revealed, these great distances can furnish a key advantage: safety from the deadly radiation that a supernova among one of our closest neighbor stars would produce.

THE DANGERS OF NEARBY SUPERNOVAE

Any supernova bright enough to stand out as a new bright star must have exploded within our own Milky Way galaxy. Even the closest galaxies to our Milky Way—the Magellanic Clouds—are so distant that a supernova within them would appear as merely a new bright star, as Supernova 1987A demonstrated. But a supernova in the Milky Way is something else. A supernova typically increases rapidly in brightness, and within a few days reaches a peak luminosity approximately 1 to 10 billion times the sun's. Then, after a few days at peak luminosity, the supernova gradually fades, dimming toward obscurity after a year or two.

The apparent brightness of any object that we see decreases in proportion to the *square* of its distance from us. But a supernova has such an enormous intrinsic luminosity that if one exploded at a distance "only" 100,000 times the sun's distance from earth, it would shine in the sky nearly as brightly as the sun does! It would also kill us in an instant with the neutrinos and the high-energy gamma rays from its explosion. Only the fish in the seas and the organisms that burrow deep beneath the Earth's surface might survive such a nearby supernova. Luckily for us, 100,000 times the sun's distance from us does not bring us even half way to the next closest star, Alpha Centauri. No supernova (always excluding our sun itself!) can appear as bright as the sun: We are in no danger of having two suns in the sky as the result of a nearby supernova.

The most likely average distance for a supernova in the Milky Way is not 100,000 times the sun's distance from Earth but 1 *billion* times the sun's distance, roughly half way across the Milky Way galaxy. At such a distance, even a supernova at its peak luminosity appears to us only about one-billionth as bright as the sun. This may sound dim, but in fact such an apparent brightness well exceeds that of Sirius, the brightest star in our night sky, by a factor of ten. In other words, a supernova that explodes at a random position in our Milky Way will, if its light is not blocked by clouds of interstellar dust, temporarily become the brightest star in the night sky.

Since supernovae explode in a galaxy such as the Milky Way once every century or so, every few generations throughout history have brought a newcomer to the constellations, a new exemplar of stellar brightness. How many of these have entered recorded history? How many have been noticed by the public at large, perhaps with trepidation, certainly with awe? The history of astronomy begins in the unrecorded recesses of time, so our knowledge of which of our ancestors saw supernovae and recognized them as something new can never be complete. But the records of supernova observations span two millennia, and have much to tell us about the stars that have exploded within our galaxy.

SUPERNOVAE IN CHINESE HISTORY

The Chinese, with the longest well-recorded history among the peoples of the world, hold the record for the oldest verified supernova. Some confusion exists, however, as to just which supernova that was, since the oldest Chinese records are fragmentary and not completely reliable. Possible supernovae exist in the records as far back as the second century B.C., but the world's experts on the historical records of supernovae, the British astronomers David Clark and F. Richard Stephenson, conclude that the "new star" seen in 185 A.D. represents the most ancient record of a supernova that is reasonably trustworthy. The Chinese chronicle covering that period states that during the second year of the reign of the Emperor Hshiao-ling, a "guest star" appeared in a certain region of the sky, which was "as large as half a mat," "showed the five colors" (i.e., was multicolored), and twinkled. The guest star faded slowly, over a period of eight months (or, on some interpretations of the chronicle, twenty months), before disappearing from view completely.

This sounds like a supernova: The twinkling confirms that the source of light was pointlike, rather than extended over a small area of the sky, like planets such as Jupiter and Venus. Because planets appear larger than the pure points of stars, when we observe them through the refracting effect of our atmosphere, planets twinkle far less than stars. The randomly changing refraction of the ray of light from each part of the planet's disk tends to cancel the refraction of the rays from the other parts.

The "guest star" of 185 A.D. had a period of fading that corresponds to the fading of most supernovae, if it did, indeed, fade over a period of eight months. Its multicolored appearance testifies to the brightness of the object, and the perceived size of "half a mat" may arise from the natural human reaction to a new, bright object. The supernova was located in the constellation Centaurus, close to being as far south in the skies as can be observed from southern China, and it barely rose above the southern horizon. As a result, the supernova's light had to pass through an especially large amount of atmosphere, and the atmospheric bending of its light would have tended to make the object seem larger than a single point—though "half a mat" seems too large to be easily explained by this analysis.

Clark and Stephenson's search of the Chinese chronicles for the years following the supernova of 185 A.D. revealed an interesting anomaly. The chronicles record possible supernovae in the years 369, 386, and 393 A.D., but these are followed by an enormous gap in time, an interval of more than 600 years, during which no likely supernovae were recorded. In the year 1006 A.D., a supernova certainly appeared, followed by another in the year 1054 A.D. (the "Crab Nebula supernova," the best studied of all historical supernova explosions), and then by a new star in the year 1181 A.D. Given that supernovae have appeared at a rate of about one per century in our Milky Way, the six-century gap that the Chinese recorded (by not recording any new stars) is statistically unlikely, but not so improbable as to be unbelievable.

After the supernova of 1181, the next new star in the Chinese records dates from 1572 A.D. This supernova—to use astronomical nomenclature, "SN 1572"—found Western astronomy in the ascendant, for that supernova, observed and recorded in detail by the Danish astronomer Tycho Brahe, played a role in persuading European minds that the heavens are not eternally unchanging. "Tycho's supernova," as astronomers sometimes rather chauvinistically refer to the explosion of 1572, was followed after a mere thirty-two years by another supernova, SN 1604, named "Kepler's supernova" after Tycho's former assistant, then successor, Johannes Kepler. The supernova of 1604 was the last star definitely *observed* in explosion in the Milky Way. However, astronomers have found the remnant of an exploded star in the constellation Cassiopeia, the famous (to them) radio source "Cassiopeia A." By measuring the rate at which matter in the remnant is expanding outward, and by extrapolating backward in time, we can date the explosion to the latter half of the seventeenth century, fifty to eighty years after Kepler's supernova. This calls for an explanation of why this explosion was not recorded in Europe, in China, or in the Islamic world. (John Flamsteed, the Astronomer Royal, apparently saw the supernova in August 1680 in England, but if so, his records of the explosion stand alone.) Before we look for such an explanation, though, we ought to take a closer look at the most important supernovae in the Milky Way, the explosions of 1006, 1054, 1572, and 1604.

THE SUPERNOVA OF 1006

Nearly a thousand years ago, at the end of April in the year 1006, the brightest recorded supernova burst forth in the constellation now known as Lupus, south of the better-known fishhook tail of Scorpio. Records of SN 1006 have survived from Japan, Korea, China, the Islamic world, and from Europe as well, where chronicles from Switzerland, France, and Italy all note the new star, although some refer to it as a comet.

The new star, which remained visible for several months, was said to "dazzle the eyes" in Europe, even though it was so far to the south as to be barely visible above the horizon. The Chinese records state that the star shone so brightly that one could see objects clearly at night by its light. Observers in Egypt said that the sky was "shining" from the light of the new star, which gave light estimated at "a little more than a quarter that of moonlight."

Using nearly twenty different records, Clark and Stephenson estimate that at its maximum light, the supernova shone with a hundred times the brightness of Venus, and (allowing for the physiology of the human eye) about one-tenth the brightness of the full moon. This maximum brightness holds the record for any supernova seen on Earth, and testifies either to a particularly luminous outburst or, more likely, to one closer to Earth than any other in the historical record. We should note, however, that even SN 1006 must have exploded at a distance of many thousands of light-years from the solar system; otherwise the supernova would indeed have outshone the moon.

THE CRAB NEBULA SUPERNOVA OF 1054

The next observed supernova, the exploding star of 1054, owes its fame not to its outburst—for it was not noticeably brighter than other supernovae that have appeared in the Milky Way—but to what it left behind, a web of gaseous filaments, still expanding from the site of an explosion, called the "Crab Nebula" after its vague resemblance to a crab.

The Crab Nebula supernova, SN 1054, was recorded in China, Japan, Korea, and the world of Islam, but *not* in Europe. In view of the European records of SN 1006, it is difficult to explain the absence of any European record of SN 1054 simply as arising from the "Dark Ages," as if the ongoing social disorder was so complete that no one wrote down what was happening. In fact, the monasteries that had recorded SN 1006 were well-organized, thriving centers of learning, and continued to be so throughout the eleventh century. Furthermore, the supernova of 1054 appeared in the constellation Taurus, easily visible high in the sky, whereas SN 1006 barely rose above the southern horizon, and yet was faithfully noted, though not correctly explained.

In contrast to the complete absence of European records of SN 1054 stands the rather full record from China, Japan, and Korea. Observers in those regions noted the appearance of a new star in the late spring of 1054—on May 27 in Japan, but only on July 4 in China; it is now believed that the Japanese date must be erroneous. The star was visible in daylight for three weeks (twenty-three days) after its first appearance, indicating that it was then as bright or brighter than the planet Venus. Thereafter it continued to fade in brightness for many months before disappearing from nighttime visibility. One of the most complete records, that of the Sung-shih (Astronomical Treatise), states:

> First year of the Chih-ho reign period, fifth month, (day) chi-chi'ou [guest star] appeared approximately [several inches] to the south-east of T'ien-kuan [the star Zeta Tauri]. After a year and more it gradually vanished.

In China, such a new star naturally attracted astrological interpretation, which indeed was the chief function of those who observed and recorded the heavens. One of the records from 1054, written by a man named Yang Wei-De, states that:

> I humbly observe that a guest star has appeared; above the star in question there is a faint glow, yellow in color. If one carefully examines the prognostications concerning the emperor, the interpretation is as follows: The fact that the guest star does not trespass against "Pi" [a stellar grouping] and its brightness is full means that there is a person of great worth. I beg that this be handed over to the Bureau of Historiography.

The grave reserve in Yang Wei-De's prognostication would do credit to any modern astrologer. This brief Chinese record lifts a veil of nearly a millennium of time to reveal a full-blown bureaucracy, ready to deal with any unusual happening.

For many years, no record of SN 1054 was known to have survived from Near Eastern sources. Then, in 1978, the astrophysicist Kenneth Brecher, working with two doctors and amateur historians, Elinor and Alfred Lieber, found a reference to it in a biography of the famous (at least locally famous) physician Ibn Butlan. Ibn Butlan was a Christian who lived and practiced medicine in Baghdad during the middle of the eleventh century, just as the great Jewish physician and theologian Moses Maimonides was to live and practice in Cordova, Fez, and Cairo during the following century. During the thirteenth century, two centuries after Ibn Butlan's life, a man named Ibn Abi Usaybia wrote a history which included material that—he stated—was copied from an account in Ibn Butlan's own hand. This account includes the following record:

> One of the well-known epidemics of our own time is that which occurred when the spectacular star appeared in Gemini in the year [1054/1055]. In the autumn of that year fourteen thousand people were buried. . . . Then, in midsummer of the [next year], the Nile was low and most people in [old Cairo] and all the strangers died, except those whom Allah willed to live. [How little things change!] . . . Thus Ptolemy's prediction came true: "Woe to the people of Egypt when one of the comets appears threateningly in Gemini!"

Brecher and the Liebers concluded that this chronicle records SN 1054, which appeared in midsummer 1054 at the boundary of the constellations Gemini and Taurus. Since the record says nothing about the new star's brightness (except "spectacular"), nor about the length of time that it could be seen, it adds nothing to our scientific store of knowledge, but the chronicle does testify to an interested and active mind that recorded a new event in the skies. Can other cultures claim as much?

Intriguingly, the only apparent surviving record of SN 1054 from what is now the "Western" world comes from the Anasazi people who occupied what is now the American southwest. Thirty years ago, the astronomer William Miller suggested that two rock paintings in Arizona might be records of SN 1054. Miller noted that on the morning after the supernova appeared, on July 5, 1054, the crescent moon happened to lie almost along the line of sight to the supernova. Any observer of the new star could hardly fail to be struck by this close conjunction of the brightest and the newest objects in the skies of night. Does it not then seem likely that the paintings of a star and crescent moon represent an attempt to memorialize this rare and stunning event?

Such conjectures can hardly be proven. On the one hand, we can date the rock paintings to the correct century (approximately). On the other hand, ethnographers have noted that the Anasazi did not usually record important *events;* the concept of a chronicle apparently had no appeal. But if only for emotional reasons, many astronomers, of whom the most enthusiastic is Jack Brandt, believe that the star and crescent moon paintings date from July 5, 1054, the sole surviving record of the supernova west of Turkey.

THE CRAB NEBULA

The supernova of 1054 owes its fame because it left behind a fascinating remnant, called the Crab Nebula. The Crab Nebula was first noted as something extremely strange during the early eighteenth century by the British astronomer John Bevis, and was listed as object number one in the list of "nebulae" compiled by Charles Messier during the final decade of the same century. (Messier, a comet hunter, wanted a list of objects that he could be sure were *not* comets; today, Messier's cometary discoveries are forgotten, but his list of pesky objects has become famous among astronomers.) A well-known remark attributed to the noted astrophysicist Geoffrey Burbidge nicely exaggerates the role that the Crab Nebula has played in astronomy: "Modern astronomical investigations are divided between studies of the Crab Nebula and everything else."

Why is the Crab Nebula so renowned? First, because it contains a "pulsar" at its center, a source of radio waves, light waves, and X rays that pulses on and off thirty times each second. This pulsar arises from the core of the star that exploded, and has become one of the most carefully studied pulsars, and for more than two decades it was the most rapid pulsar yet discovered. Second, the Crab Nebula itself offers us almost the nearest, and by far the most carefully studied, remnant of the outer layers of an exploded star. This remnant consists, in part, of an expanding web of gaseous filaments made of hydrogen gas that testify to an outburst in astronomically recent times (Figure 11). These filaments are embedded within a more evenly spread-out sea of light, produced by the process called "synchrotron emission," which occurs only when particles have been accelerated to speeds close to the speed of light by the rapidly rotating stellar core (see Chapter 11).

Thus the Crab Nebula furnishes astronomers with a supernova remnant seen less than a thousand years after its explosion, at a distance of about 5,500 light-years—about one-thirtieth of the distance to Supernova 1987A. During the next millennium, astronomers plan to observe SN 1987A as carefully as they can, to see how it resembles the Crab Nebula in its development, and how it differs. From such similarities and differences grow advances in our knowledge of exploding stars.

TYCHO BRAHE'S SUPERNOVA

On the evening of November 11, 1572, the Danish astronomer Tycho Brahe was walking homeward, contemplating the sky, when

> [B]ehold, directly overhead, a certain strange star was suddenly seen . . . Amazed, and as if astonished and stupefied, I stood still.

Tycho reacted just as Ian Shelton would more than four centuries later:

FIGURE 11. The Crab Nebula consists of material spewn forth from the exploding star of 1054, which we now observe nine and a half centuries after the explosion. (Palomar Observatory photograph)

> When I had satisfied myself that no star of that kind had ever shone forth before, I was led into such perplexity by the unbelievability of the thing that I began to doubt the faith of my own eyes, and so, turning to the servants who were accompanying me, I asked them whether they too could see a certain extremely bright star when I pointed out the place directly overhead. They immediately replied with one voice that they saw it completely and that it was extremely bright. . . . And at length, having confirmed that my vision was not deceiving me . . . I got ready my instrument. I began to measure [the star's] situation and distance from the neighboring stars of Cassiopeia, and to note extremely diligently those things which were visible to the eye concerning its apparent size, form, color, and other aspects.

Tycho had detected the first supernova to appear in the Milky Way in four centuries. The Renaissance had brought a new world vision, first to southern and then to northern Europe, a view of nature exemplified by Tycho's "scientific" attitude, ready to observe as well as he could the new phenomenon that nature had brought forth. Brahe's ancestors, proud lords of a Danish fen, would doubtless have seen the hand of God and little more—and would hardly have left behind a detailed record for posterity.

But Tycho Brahe was the man who, almost single-handedly, stumbled onto the bedrock necessity of modern science: accurate, repeated observations, not a chronicle based on preconceptions of what *ought* to occur but instead a record—subject to such bias as can not be eliminated from the process—of what *does* occur. This remains the scientific ideal, without which all scientific hypothesizing turns into so much hot air, for without accurate data, the clash of theories becomes meaningless.

Tycho (for historical reasons usually his first name is used) was only twenty-five years old when SN 1572 appeared. He had been sent to the universities at Copenhagen and at Leipzig to study law, but he had already shown a strange propensity for the study of nature. When Tycho acquired two books of astronomical tables that predicted the positions of the stars and planets, he discovered to his amazement that the tables were noticeably incorrect. There is hardly anything as stimulating to a young scientist as the discovery of another's error; it simultaneously gives confidence in one's own work and a chance to compete in the big time. After receiving an inheritance from his uncle, Tycho traveled to Germany to study science. There, following an argument with another student about a mathematics problem (how seriously they took mathematics in those days!), he fought a duel with swords and lost part of his nose, apparently along the bridge. Tycho could afterward be recognized by the silver prosthesis fitted into his reshaped proboscis, although the silver was painted so as "to look real."

The sudden appearance of the supernova in 1572 changed Tycho's life: From then on it was astronomy and nothing else. He wrote a book on the new star, published in 1573, which, quite typically for the age, opened with an astrological interpretation, quite like the Chinese records or those of Ibn Butlan. But in addition Tycho recorded, carefully and accurately, the position of the new star and the changes in its brightness. Hence SN 1572 is the first supernova for which we have an accurate "light curve," the time history of the object's apparent brightness. Since Tycho, astronomers have attempted to secure a carefully recorded light curve for every supernova, and these light curves have proven a key tool in understanding stellar explosions. For example, Type I and Type II supernovae differ noticeably in their light curves—a difference that arises from their different explosion mechanisms.

The fame that Tycho gained through this book brought him to the attention of the Danish king, who already knew Tycho's family as an important one. The king gave Tycho the island of Ven, in the sound between what are now Denmark and Sweden, where Tycho built an observatory, his "Uraniborg" (castle of the heavens). Here, even before telescopes were invented, Tycho assembled the instruments that allowed him to survey the skies more accurately than ever before. Most notable among these was a giant transit, a long rod mounted on a north–south wall, which allowed careful measurement of the altitude of any star—its height above the horizon—as it crossed the meridian, the imaginary line joining the north and south points on the horizon through the zenith or overhead point (Figure 12).

Tycho's genius lay in his observations, which he continued through the last two decades of the sixteenth century. After the Danish king who had so befriended Tycho died, Tycho found it necessary to relocate. After some hesitations, in 1599 he arrived

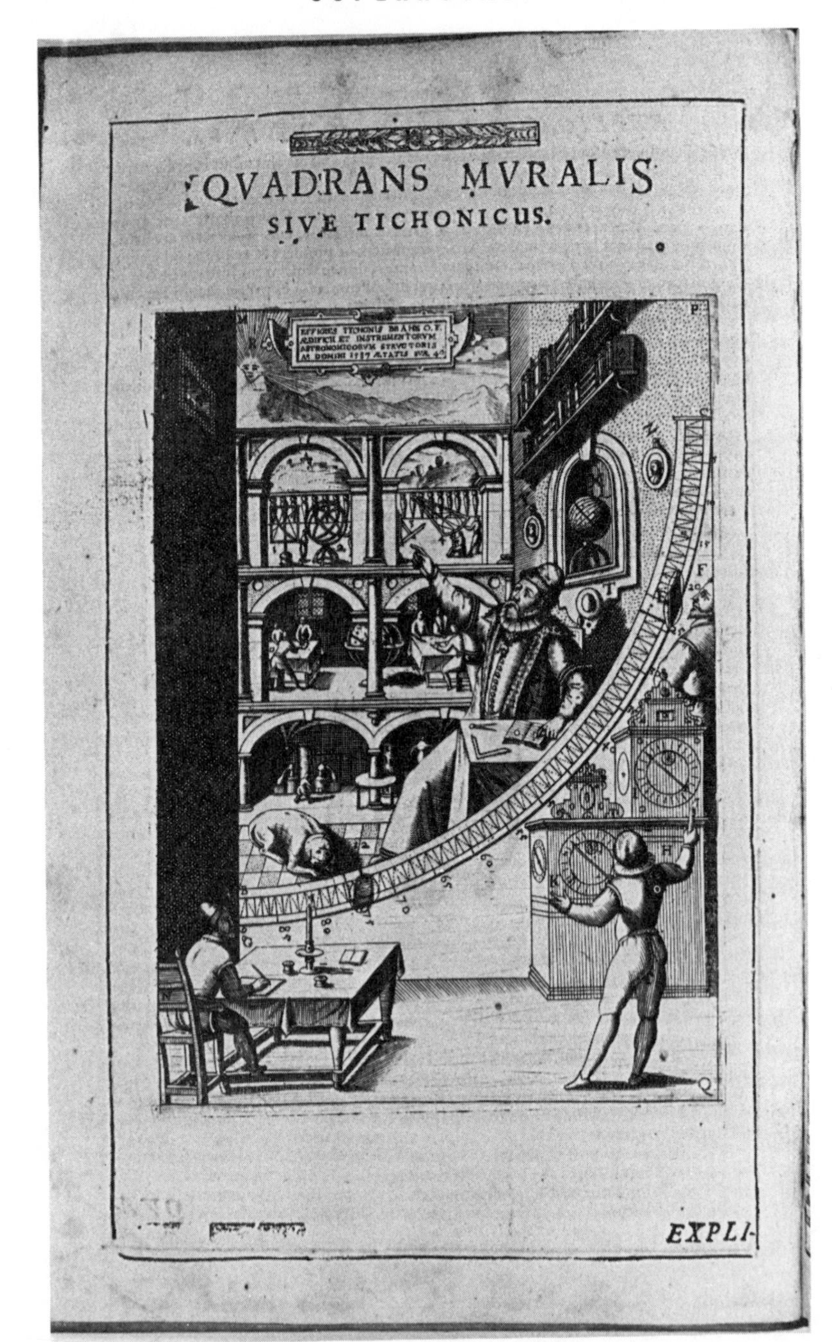
QVADRANS MVRALIS
SIVE TICHONICUS.
EXPLI-

in Prague, at the court of the Holy Roman Emperor Rudolf. There, in the fall of 1601, Tycho attended a formal banquet at which, apparently in the throes of etiquette, he neglected his human needs with fatal effect. His death from a bladder infection followed soon thereafter. Conveniently for astronomy, however, Tycho had by that time engaged (with far too little respect) an assistant whose mathematical abilities far outshone his own, and who would soon determine the correct model for the solar system: Johannes Kepler.

KEPLER AND THE ORBITS OF PLANETS

Kepler came from the lower classes and owed his advancement to his intellect, which had been recognized and nurtured by his astronomy instructor at the University of Tuebingen, Michael Maestlin (who, however, could not go along with Kepler's immediate acceptance of the newfangled Copernican model of the solar system). Kepler became a schoolmaster in Graz, on the fringes of the empire, where he wrote a slim book speculating on the reasons why only six planets should exist, and why the planets' orbits should have the relative sizes that we observe.

This book, although entirely wrong in its major conclusions, showed a keen and fertile mind; when Tycho read it, three years before he died, he recognized a useful future assistant. Invited by Tycho, Kepler moved to Prague, only to discover that his meager salary generally went unpaid, that he was treated more like a servant than a coworker, and that Tycho would not reveal the treasure trove of data gathered from more than twenty years of observation.

Tycho's death brought an immediate improvement in Kepler's fortunes: The emperor invited him to take over as Imperial

---

FIGURE 12. This sixteenth-century engraving depicts Tycho Brahe in his observatory, Uraniborg, directing his assistants at work. The great quadrant measured the angle above the horizon of stars and planets as they crossed the meridian, the line connecting north and south on the horizon and passing through the point overhead. (Owen Gingerich)

Mathematician. Now Kepler had a princely salary (which unfortunately also languished unpaid), he received some respect on the rare occasions when he dared to appear at court, and—by far the best—he had complete control over the greatest set of astronomical data in the world. Kepler set out to determine the orbit of the planet Mars, an effort that took a few years longer than he had originally hoped. He completed his task in 1609, when he happily announced that Mars orbits the sun along an elliptical trajectory with the sun at one focus of the ellipse. This work led directly to the final acceptance of the Copernican model and, later, to Isaac Newton's triumphant demonstration that the planets' elliptical orbits follow naturally from Newton's law of gravitation. But before this all happened, another unexpected event intervened: Kepler's supernova.

THE SUPERNOVA OF 1604

Kepler suffered from poor eyesight, no doubt worsened by his love of reading and calculating, but he nevertheless has a supernova named (among astronomers) in his honor, SN 1604. This supernova, like SN 1572, was observed in Japan, China, and Korea, where records help to establish the date on which the new star first appeared, and the rate at which its brightness diminished. But Kepler deserves primary recognition because, carrying on the tradition that Tycho began, he made repeated, careful measurements of the supernova, impelled by the desire to record all that could be observed.

This modern attitude now seems too familiar to merit comment, but, as the scattered records from the Orient show, there was a time and a place—namely, the late sixteenth century in Europe—where this approach first flourished, bearing fruit in succeeding centuries in a thousand different discoveries. All scientific discoveries spring from the systematic accumulation of data, without which the most unanticipated events—a new star, for example—cannot be recognized, let alone correctly interpreted.

The new star of 1604 was first recorded on October 9 of that year by two observers in Italy, one in Verona and another in Cosenza, who reported his discovery to Father Clavius, a Jesuit astronomer in Rome now best known for his controversies with

Galileo. On the next night the star was recorded in Padua, and also in Prague. This naturally led to a report to the Imperial Mathematician, and Kepler sprang into action, collecting reports from all over Europe and making repeated observations of the supernova throughout the year 1605.

From late October 1604 until January 1605, we find no reliable observations from Europe, and few from the Orient, because the Earth's motion around the sun placed the sun more or less along the line of sight to the supernova. This prevented a good view of the new star, which was located in the constellation Ophiuchus, relatively close to the ecliptic (the sun's apparent path around the sky through the twelve constellations that form the zodiac). In January 1605, after the supernova became visible again, it was first brighter than Antares (the brightest star in Scorpio), then slightly less bright (late February), then noticeably less bright. It then faded into invisibility, last seen on October 8, 1605, when Kepler wrote that "Now exactly a year after its first apparition, in a very clear sky, its appearance could be noted only with difficulty."

Tycho's supernova, in contrast to Kepler's, had a position not only far from the ecliptic, in Cassiopeia, but so close to the north celestial pole—the point on the sky directly above the Earth's north pole—that the new star never set for observers in northern Europe. (Analogously, Supernova 1987A never sets for observers located in Chile.) But then, Tycho was a lucky man, whereas Kepler had a far more difficult life, which included expulsion from the Catholic region near Graz for being a Protestant, and having to defend his mother (successfully) against charges of witchcraft. Despite these and many other difficulties, Kepler persevered to find the shapes of the planets' orbits around the sun, as well as the laws that connect their speeds in orbit and their distances from the sun. Kepler died, during the Thirty Years War that ravaged Germany, on a futile journey to recover money owed him by a publisher—he should be the patron saint of authors!—and his grave has been obliterated for three and a half centuries. But his fame lives on, and includes the last supernova to be well observed in the Milky Way galaxy.

THE HIDDEN SUPERNOVA

Technically, Kepler's supernova was the next-to-the-most recent supernova to be seen in our galaxy. In the constellation Cassiopeia, only a few degrees from the spot where Tycho's supernova appeared in 1572, a web of expanding filaments of gas testifies to a more recent explosion. In addition to this evidence, astronomers now have the ability to detect radio waves emitted by cosmic objects, typically objects such as supernovae, which have recently undergone some type of violent outburst. Radio observations of the region in Cassiopeia show significant amounts of radio waves that were apparently produced by the process called "synchrotron emission," a tip-off to a supernova explosion. The astronomers who have studied the web of gas, and the spectrum of radio waves from it, believe that this region, called Cassiopeia A (Cas A for short), has the look of a supernova remnant (Figure 13). This remnant's radio emission resembles the radio waves emitted at the locations where Tycho's and Kepler's supernovae appeared and then disappeared.

But there's a problem. Cas A emits radio waves much more intensely than either of the two known supernova remnants. This implies that if Cas A arose from a supernova explosion, the explosion must have been intense, even for a supernova, and must have occurred recently, because as a supernova remnant expands over time, its radio emission eventually weakens progressively. In addition, and even more convincing to astronomers, the speed at which material is expanding in Cas A's gaseous filaments can be measured. If we combine this speed with our estimate of Cas A's size, which we obtain from an estimate of its distance and a measurement of its angular size on the sky, then we can extrapolate back in time to discover how long Cas A would have had to expand at its present rate in order to achieve the size that we now observe.

This answer turns out to be that we observe Cas A a bit more than three centuries after its explosion. Since the filaments in Cas A are unlikely to have speeded up their expansion, but in fact are more likely to have slowed somewhat as time has passed, our best estimate for the time that the supernova appeared is just about 300 years before the present; that is, in the late seventeenth century. And this seems odd: Science, and in particular

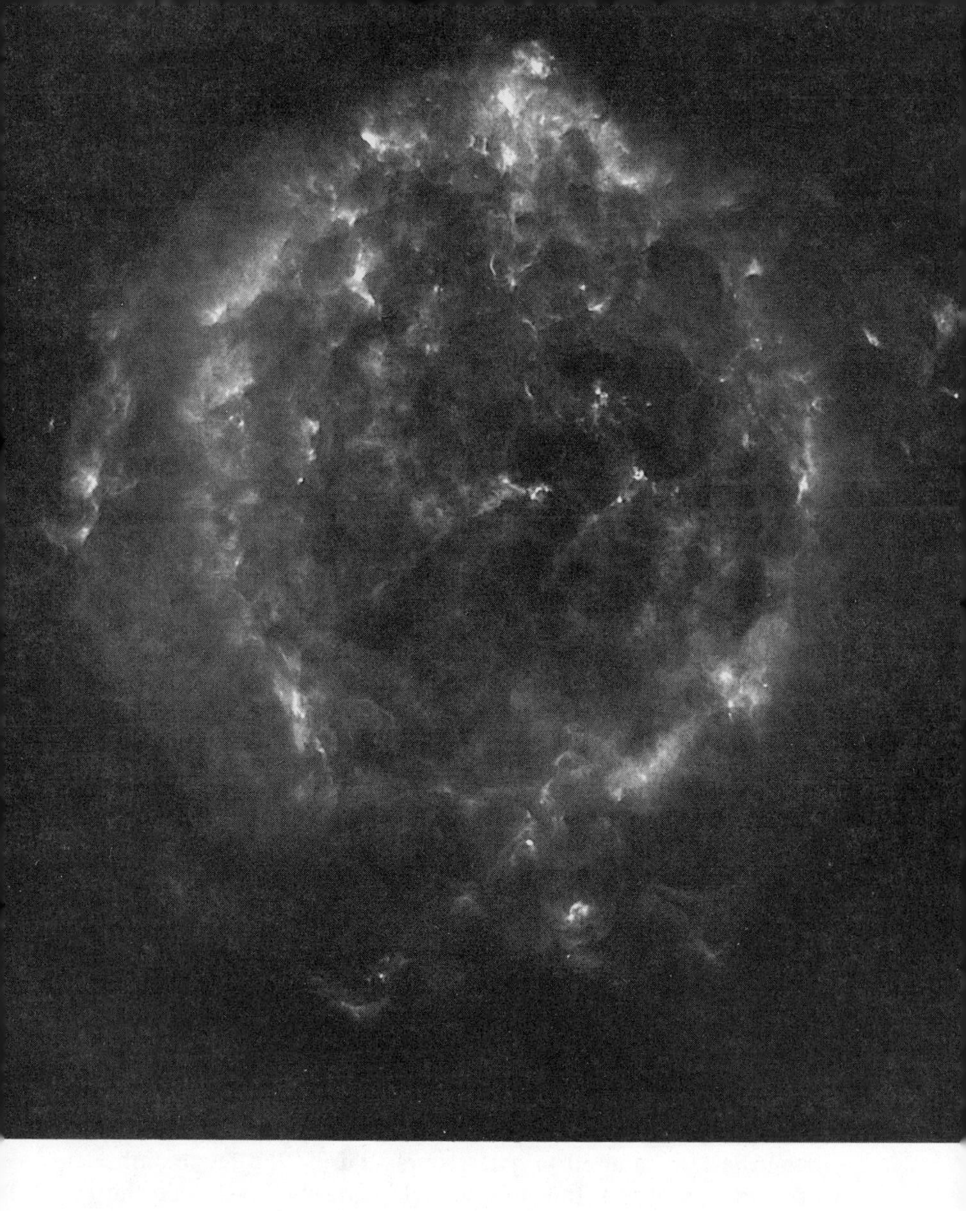

FIGURE 13. This "photograph" records the radio waves emitted by the supernova remnant Cassiopeia A. The radio emission forms a (roughly) spherical shell, which is expanding away from what is believed to be the point where a supernova exploded. (National Radio Astronomy Observatory)

astronomy, was then flourishing in Europe, not to mention the traditional sites of supernova observation—Japan, China, and Korea—or the newly founded colonies in America. How could a supernova pass unnoticed by Isaac Newton, Edmund Halley, Christian Huygens, and a host of other astronomers, all experts at observing the sky and at improving telescopes? Did a veil fall on the world of scientific observation during its most flourishing period?

Apparently just this did occur, though the blindness arose not on Earth but in interstellar space. Along the line of sight to Cas A lies a region of high "interstellar absorption," the dimming of starlight caused by interstellar dust particles. The dust grains made when interstellar atoms collide and stick together lie scattered through interstellar space. Each of these grains can block or "absorb" some of the starlight, reducing the amount that penetrates the dusty region. You can see this absorption by interstellar dust on a clear fall night, if you look toward the west at the "milky way," a pale band of light in the constellations Cygnus and Aquila composed of millions of stars that concentrate in the disk of our galaxy. Within Cygnus, a particularly large amount of interstellar dust blocks the light from the central plane of the milky way, creating an optical-illusion effect in which two separate bands of light appear to exist. Between these two bands lies interstellar dust, detectable not by its own emission of light but through the fact that it *absorbs* the light from behind it.

The interstellar absorption in the direction of Cas A is so extreme that it can make a supernova look like just another star! To be sure, the supernova itself remains unaffected; the absorption simply prevents most of the light headed in *our* direction from reaching us. The supernova that produced Cas A could have had its light so attenuated that it would have seemed no brighter than a second- or third-magnitude star—that is, just one among the few hundred brightest stars—during the few months when it was brightest. It is also possible that, like SN 1987A, the Cas A supernova had an exceptionally low intrinsic luminosity. This would help to explain why keen-eyed astronomers missed it, even though they were assiduously studying the stars in detail. Perhaps winter weather, a low-luminosity explosion, a temporary lull in attentiveness, and sheer bad luck all combined to prevent discovery (except, as noted, by John Flamsteed) of the much-attenuated light from the new star. Or perhaps today's astronomers have incorrectly analyzed Cas A, and it is not a supernova remnant, or not a remnant so young as it appears to be.

HOW MANY MORE SUPERNOVAE ARE HIDDEN BY INTERSTELLAR DUST?

If the late-seventeenth-century supernova that produced the Cas A supernova remnant was never noticed in the skies of Earth, does this not imply that other supernovae may well have exploded unseen in our own galaxy, flared out and then dimmed into obscurity without human notice? Indeed it does, and astronomers even estimate that as many as half, perhaps even more than half, of all the stars that explode in the Milky Way are likely to remain invisible to human eyes. Absorption of light by interstellar dust plays a more significant role in our view of the universe than is generally realized; here we have a prime example of the obscuration from interstellar dust that prevents our enjoyment of one of nature's grandest spectacles.

A spiral galaxy such as our Milky Way has a shape like a discus, much thinner in one dimension than in the other two (see Figure 3). Interstellar dust particles concentrate heavily toward the galactic plane of symmetery, that is, toward the imaginary plane that divides the "top" and "bottom" halves of the galactic discus. Unfortunately for supernova observations, the stars that produce most of the supernova explosions—the "Type II supernovae," the explosions of massive, aged stars, which include SN 1987A—likewise concentrate toward the plane of the Milky Way. Hence interstellar dust finds itself well situated to absorb the light from a Type II supernova, and this accounts for the fact that half or more of all such supernovae may have passed undetected on Earth.

Relying on what they know about interstellar dust, and about supernova explosions observed both in the Milky Way and in other galaxies, astronomers estimate that a supernova explodes in a large galaxy such as our own about once every fifty years. Some astronomers would substitute once every thirty years for fifty; others lean toward once every hundred years. We can round off the figure mercilessly (always a good idea with astronomical facts), and use a figure of one supernova explosion per century in a large galaxy. This implies, of course, that we are "overdue" for another supernova in the Milky Way, but guessing that a supernova will therefore soon be seen is no more certain than betting on the red at roulette when black has come up four times running: The present chances are unaffected by the past.

With supernovae, however, the past *is* the future, in the sense that we observe only the past. The next few hundred supernovae to be detected on Earth have already exploded; if they are located within the Milky Way, they probably exploded many thousands of years ago, and their light has been piercing space at 6 trillion miles per year ever since. Soon—we don't know just when—some of that light will reach the Earth. Then the news will spread that we have seen a supernova not simply in our closest neighbor galaxy, but within our own Milky Way.

SUPERNOVA IN ANDROMEDA

A taste of what we might someday see within our own galaxy appeared in our closest galactic twin, the Andromeda galaxy, in 1885. The Andromeda galaxy and our Milky Way are by far the two largest and most massive galaxies in the Local Group, our small cluster of galaxies. Each of these two giant spirals spans a diameter of at least a 100,000 light-years and contains several hundred billion stars. And each galaxy has two good-sized satellite galaxies—the Magellanic Clouds for the Milky Way, and two elliptical galaxies for Andromeda. Each giant spiral also produces a supernova once every century or so. In the case of the Andromeda galaxy, we have seen one and only one supernova, that of 1885.

Since the Andromeda galaxy lies at a distance of 2 million light-years from the Milky Way, it is hardly surprising that we never saw a supernova until the late nineteenth century; before then, our telescopic exploration was too sporadic and too low-powered to reveal a supernova clearly. But in 1885, no doubt existed that the Andromeda Nebula, as it was then called (for astronomers had no clear picture of the distribution of matter into galaxies), had produced a new object that shone with about one-tenth of the light from the entire galaxy!

Unfortunately, the light from the supernova, which had traveled for 2 million years, arrived about five years too soon, just before the technique of astronomical spectroscopy—of dividing starlight into colors and studying those colors one by one—reached the point where we could have recorded the details of what the supernova was made from, and how rapidly its outer layers were expanding. We may therefore note with simple regret that SN 1885 has little to teach us; in many ways we have more to learn from much earlier supernovae such as SN 1006, SN 1054, and SN 1572 within our own galaxy. Supernova 1987A, the first exploding star to be seen in our Local Group of galaxies since SN 1885, therefore represents an opportunity to be seized and made the most of. In order for us to do so, we must take the time to understand how stars are born, age, and die, most in quiet despair, a few, like SN 1987A, in spectacular glory.

6

THE LIVES OF THE STARS

IN THE heavens as on Earth, the universal rule of "ashes to ashes, dust to dust" governs the lives of the stars. All around us we see shining examples of stars in full glory, no longer surrounded by the clouds of gas and dust within which they were born, but not yet dying, not yet devoid of the ability to release energy that radiates through the universe. Whence comes this ability? What provides the radiant energy of the universe, the starlight that shines day and night?

HOW STARS SHINE: NUCLEAR FUSION

All stars shine because of high-energy collisions that make atomic nuclei *fuse*—stick together to form a new type of nucleus. Such "nuclear fusion," the melding of two atomic nuclei into a single, larger nucleus, makes all the normal stars of the universe shine. Within stars, nuclear fusion melds hydrogen nuclei—individual protons—into helium nuclei, trillions upon trillions of times per second, day after day, year after year, millennium after millennium. In nuclear fusion lies the secret of starlight—a secret hidden within the cores of the stars, deep below the stars' visible surfaces.

Stars consist of atomic nuclei—chiefly protons and helium nuclei—plus electrons. All of these particles are in constant, seething motion, with temperatures of many thousands or millions of degrees. The nuclei can undergo nuclear fusion; the electrons cannot. Since the 1930s, scientists have known that if atomic nuclei collide so violently that they fuse together, their fusion produces heat and light.

This heat and light arises from the fact that nuclear fusion causes nuclei to *lose mass.* Part of the mass, which provides a measure of how much matter the nuclei contain, vanishes. But the mass does leave a trace; it is *transformed* into the heat and light that stars produce. Nuclear fusion is the greatest magic act the universe has to offer, the means by which apparently dull and impotent matter can yield a universal source of energy. This is the secret of the stars, and of hydrogen bombs as well: If you can throw nuclei at each other in such a way that they fuse together, you can turn part of their mass into heat and light, which you can use to blow down a city—or light up the universe.

SCIENTIFIC DEFINITIONS OF ENERGY AND MASS

If you want to understand how stars turn mass into energy, you have to begin with an accurate definition of what these terms mean. To a scientist, "energy" measures the capacity to do "work." Work is measured as the product of the amount of "force" exerted on an object, times the distance over which the force acts. And "force" is whatever produces an acceleration—i.e., a change in an object's *speed* of motion, or in its *direction* of motion. Hence with more energy, you can keep on applying more force to an object over a greater distance.

If you exert force on an object—say, by swinging a golf club at a golf ball—you can accelerate the object. The amount of acceleration will depend both on the amount of force and on the *mass* of the object to be accelerated by that force, in this case the golf ball. The mass of an object measures the quantity of matter that it contains. This mass does not change so long as the object remains unchanged. If we send the mass into space, we may change the gravitational force exerted upon it (what we call the object's "weight" at the Earth's surface), but we do not change the object's mass unless, for example, we break the object into two or more pieces. We can measure an object's mass by its *resistance* to being accelerated by a given amount of force: More massive objects are harder to accelerate. Thus, if you swing a golf club at a softball, you will not see the ball leap off the tee in the same way that a golf ball does, because the softball has more mass.

For any particular object, a greater amount of force exerted on the object will produce a greater acceleration. If you seek more rapid acceleration in your automobile, you need a more powerful engine, in order to create more energy per second and thus to exert more force on your car. If you want to accelerate objects of different masses at the same rate, you need a greater force for the more massive object. Thus, for example, a railroad locomotive's engine produces far more energy per second than the engine of a semitrailer in order to accelerate the railroad train at about the same rate as the acceleration produced by the semitrailer's engine.

ENERGY OF MOTION AND ENERGY OF MASS

Energy comes in various forms, of which the most obvious (to us!) is *energy of motion* or kinetic energy—the energy possessed by particles in motion by virtue of their motion. For those objects with mass (anything except massless particles such as the photons that form light waves), an object's amount of energy of motion varies in proportion to its mass times the square of its velocity. As we would expect, objects moving at greater speed have more energy of motion: An automobile speeding at 60 miles per hour has four times the kinetic energy of one moving at 30 miles per hour. Likewise, a semitrailer moving at 60 miles per hour has far more energy of motion than an automobile traveling at the same speed, simply because it has far more mass than the automobile.

The other key form of energy is *energy of mass,* the energy that resides in any object with mass. Locked within any amount of mass, in existence simply because that mass exists, is a corresponding amount of energy of mass, an amount given by Einstein's most famous equation:

$$E = mc^2$$

Because c, the speed of light, is an enormous number, so too is the energy of mass, E, contained in any modest amount of mass m. The energy of mass contained in a particle is as basic to the existence of that particle as any property more appealing to the senses—height, depth, weight. You can't feel it or taste it, but you've got it, simply because you have mass. You and I carry our energy of mass easily, even uncaringly, unaware (most of the time) of the great potential resource we possess. Einstein's formula allows us to calculate that even a quarter has enough energy of mass to supply the energy for New York City during an evening rush hour. The trick is to transform the energy of mass in this quarter into a form of energy useful to us—energy of motion, the type of energy that powers electrical generators or moves automobiles. As to the energy of mass in a human being, 10,000 times greater than that in a quarter, the conversion to "useful" energy is dreadful to contemplate, perhaps unnecessary—so long as we have stars. (To perform this conversion, the first step is to heat the matter to about 20 million degrees, so that nuclear fusion can occur.)

TO LIGHT THE UNIVERSE WITH NUCLEAR FUSION

As mentioned above, the key to producing energy through nuclear fusion is that when nuclei fuse together, *mass disappears:* The particles that emerge from the process have a total mass that is slightly less than the mass of the particles that collide to begin the fusion reaction. The details of this fusion deserve examination, for they produce all of what we call natural light, and are the key to understanding what produces a supernova.

Deep inside stars, nuclear fusion transforms energy of mass into energy of motion. This transformation takes place at the stars' centers, the only places hot enough for nuclear fusion to occur. As a result of nuclear fusion, the total mass in the star decreases slightly. So too does the amount of energy of mass, but the vanished energy of mass is *replaced* by an exactly equal amount of energy of motion. Energy has the appealing property that throughout the universe it can be neither created nor destroyed, but only changed from one form (energy of mass, for example) into another (energy of motion).

In most stars, nuclear fusion proceeds through three steps: First, two protons fuse together to make a larger nucleus called hydrogen-2. Next, a proton fuses with a hydrogen-2 nucleus to make a nucleus of helium-3. Finally, two nuclei of helium-3 fuse together to make a nucleus of helium-4 (ordinary helium) and two protons. In each of these fusion reactions, the particles that emerge from the fusion have a total mass that is slightly less than the sum of the masses of the individual particles that entered the fusion reaction: Some energy of mass has been transformed into energy of motion.

If you want to produce nuclear fusion in this way, all you need is a good supply of protons and a temperature between 15 and 60 million degrees Fahrenheit. At lower temperatures, nuclei will repel each other because they each carry a positive electric charge, and like electric charges tend to repel one another. Only at temperatures of many millions of degrees will the nuclei move rapidly enough to overcome their mutual repulsion and fuse together. Hence nuclear fusion can hardly be a low-temperature process—which is why we don't have any in our daily lives.

On Earth, working with immense ingenuity, humans have managed to mimic the process by which stars produce energy of motion by creating "thermonuclear weapons," also known as "hydrogen bombs." Like stars, hydrogen bombs fuse hydrogen into helium nuclei at enormous temperatures—temperatures that we produce momentarily by exploding an "atomic bomb" that produces energy of motion from the disintegration of rare nuclei of radioactive uranium or plutonium.

Earth's most destructive hydrogen bombs fuse a few pounds of hydrogen into helium. But every second, the sun's core detonates the equivalent of 100 billion hydrogen bombs, fusing an enormous mountain's worth of hydrogen into helium nuclei. When we attempt to produce energy of motion by duplicating this process in "controlled nuclear fusion," we face enormous difficulties in confining matter at temperatures of 20 or 30 million degrees Fahrenheit, and in controlling the energy of motion that emerges. Of course, a hydrogen bomb makes no such attempt: For a microsecond, the new energy of motion bursts forth, unconfined and uncontrolled.

How does the sun control its enormous output of energy of motion without bursting apart, and thus succeed where we have failed? The sun contains an enormous amount of material—about 330,000 times as much mass as the Earth does. This gaseous mass forms a barrier to the nuclear fusion fury at the sun's center. If we look for "controlled nuclear fusion," we need look no farther than the sun and its sister stars, which wrap their nuclear fusion in enormous protective blankets of matter unavailable to us on Earth.

FROM THE CENTER OUT: THE TRANSFER OF ENERGY WITHIN STARS

In the core of every star, countless times per second, the energy of motion made from energy of mass via nuclear fusion appears in the form of *additional velocity* acquired by each of the particles that emerge from the nuclear fusion. Before the fusion, the particles had energy of motion, since each of them was in motion. After the fusion, the particles have *more* energy of motion, which they gained from the energy of mass that vanished during the fusion process.

What happens to the energy of motion of the particles that emerge from the fusion? The particles collide with particles immediately around them, which in turn collide with other particles, and they in turn collide with still others, until the newly fused particles share the energy of motion made at the star's center with the entire star. Likewise, high-energy photons made during the nuclear fusion collide with other particles and increase the particles' velocities. Eventually, like a mob animated by a demagogue at its center, all the particles within the star dance in a frenzy induced by the nuclear fusion at the stellar core. The particles dance most furiously at the star's center, progressively less so in the outer regions. Thus, from the nuclear fusion at its core, the entire star grows hot from center to surface. The star's center typically has a temperature of 15 to 60 *million* degrees Fahrenheit, while the surface temperature falls to a mere 2,000 to 25,000 degrees.

Because a star is *hot,* it produces electromagnetic waves. Any object not at a temperature of absolute zero—the coldest temperature possible, −459.67 degrees Fahrenheit—produces electromagnetic radiation, more and more of it as the object grows hotter and hotter. Furthermore, as an object grows hotter, the chief *type* of electromagnetic waves it radiates will change. Human beings and other objects near room temperature produce mostly infrared waves. Hence the sizable military industry that has sprung up to detect the infrared waves that humans emit, in order to "see" the enemy simply by the waves that no human can avoid radiating into space.

At temperatures of a few thousand degrees, an object will emit mostly visible light. Stars, with surface temperatures measured in thousands of degrees, therefore radiate mostly visible light, along with sizable amounts of ultraviolet from the hotter stars. Not accidentally, our eyes have evolved to detect mostly visible light, the sun's primary output.

In the core of a star such as the sun, where the temperature rises to millions of degrees, the hot gas radiates mostly X rays and gamma rays. If the outer layers of the sun were transparent to this radiation, we would be instantaneously "zapped" by these high-energy photons from the solar interior. However, the matter in the sun effectively traps all the X rays and gamma rays, each of which travels only a tiny distance within the sun before encountering a nucleus or an electron which blocks its path and deflects it in another direction. The high-energy photons therefore cannot escape from the sun; instead, their energy is constantly passed to the other particles within the sun, heating them still further. The immense number of collisions slowly lessens the energy of each photon, and if we could pass outward in the sun from its center to its surface, we would find mostly gamma rays and X rays at the center, mostly X rays and ultraviolet in its middle regions, and mostly ultraviolet and visible light near the surface. Finally, from the regions close to the sun's surface, photons *can* escape, but because these regions have temperatures of only about 10,000 degrees Fahrenheit, the photons that do escape are ultraviolet and visible-light photons, the kind that matter at 10,000 degrees produces.

THE BENEFITS OF NUCLEAR FUSION

We on Earth live parasitically on the sun's nuclear fusion, though our planet intercepts about one part in a billion of the sun's energy output. This amount suffices to supply the energy to all forms of life on Earth; among the few exceptions are the recently discovered tube worms that live in the deep-sea vents in the Pacific Ocean, feasting on the heat from the Earth's interior, which arises from the decay of rocks that contain radioactive nuclei such as uranium and thorium.

Plants use the sun's energy to grow by constructing molecules of carbohydrates from the matter in the soil, the atmosphere, and water. Since only about one part in a thousand of all the solar energy that reaches the Earth goes into such photosynthesis, we could theoretically grow a thousand times more photosynthesizing plants than we do now, making our planet far different. Meanwhile, animals on Earth eat plants, or eat animals that eat plants, or (in extreme cases) eat animals that eat animals that eat plants. The entire ecological system on Earth therefore runs on solar energy—energy of motion that appears deep in the solar interior, passes to the sun's surface through countless collisions, and then leaps through 93 million miles of interplanetary space in about eight minutes' time—our eight additional minutes in the event that the sun should fail to shine.

THE BIRTH OF STARS

Every star that shines, and every supernova that explodes, began its stellar life inside a vast interstellar cloud of gas and dust (Figure 14). Within such a cloud, individual clumps of gas began to shrink. As they contracted, each of these clumps grew hotter as it became denser, because the fundamental laws of physics tell us that any collection of atoms will raise its temperature when it is compressed into a smaller volume. The contraction, slow at first, simply heated the gas in the cloud to temperatures of a few hundred degrees. But as the cloud shrank to a still smaller size, the temperature rose to thousands of degrees. Now the atoms were moving so rapidly that when they collided, they knocked the electrons loose from the nuclei, until the gas contained only free-roaming electrons, none of which was in orbit around the nuclei. Eventually, after several hundred million years of contraction, the temperature at the center of the clump rose to tens of millions of degrees. At this point, the hydrogen nuclei at the center began to fuse, and a star was born from what had been a contracting "protostar," a star in the formation process.

WHY DON'T ALL STARS EXPLODE—OR COLLAPSE?

If nuclear fusion did not exist, a contracting protostar might contract forever. Nuclear fusion, however, provides the means for a star like our sun to cease its contraction, and to maintain a constant size and a constant rate of energy output over billions of years.

To see why this is so, consider the fact that within interstellar

clouds, clumps of gas and dust contract because of their *self-gravitation,* the gravitational force that each part of the clump exerts on all the other parts. In an object with the mass of a star, the gigantic amount of matter produces a correspondingly gigantic amount of self-gravitation, and this mutual attraction among all parts of the object creates an enormous compressive effect. If this were the entire story, a star would quickly become a "black hole," an object of near-infinite density and nearly infinitesimal size—and we would not be the happy children of solar energy.

But within every star, another process opposes the tendency to contract: nuclear fusion. Because nuclear fusion turns energy of mass into energy of motion, the star continuously creates new energy of motion, which diffuses outward through collisions among all the particles in the star. The fact that the particles inside the star are *hot,* and therefore dance at high velocities, *opposes* the star's tendency to collapse under its self-gravitation. So long as the star can replace the energy that diffuses outward with new energy from nuclear fusion, it can maintain a constant size. If you picture the center of a star, correctly enough, as a place where a hundred billion hydrogen bombs are exploding each second, it is easy to see that the star has a tendency to explode, which indeed it would do if gravity disappeared. Conversely, if nuclear fusion ceased, even for a few seconds, the star would collapse under its own gravitation. Every star that shines represents the most perfect and fundamental balance, a natural, self-imposed matching of the star's nuclear fusion and self-gravitation.

THE IMPORTANCE OF TEMPERATURE AT A STAR'S CENTER

Consider a star that is turning energy of mass—Einstein's mc^2—into energy of motion at its center. In a typical star, each second

---

FIGURE 14. A stellar nursery such as the Eagle Nebula will eventually (in a few million years) produce a cluster of several thousand stars. The first-born of these stars already light the nebula from within; the darkest regions are the most likely sites for stellar birth in the near future. (National Optical Astronomy Observatories)

involves the conversion into kinetic energy of the mass of several mountains, and the production of enough energy of motion to last our civilization for a billion years. The nuclear fusion that produces this result is tremendously sensitive to *temperature:* If the temperature at a star's center were to double, the rate of energy production would increase nearly twenty times. At higher core temperatures, the rate of nuclear fusion becomes even more sensitive to changes in the temperature. This sensitivity arises from the fact that in order to fuse, the nuclei must overcome their mutual repulsion, caused by the fact that they each have a positive electric charge. The nuclei can do so only by moving with extreme speed, and the hotter the star's center becomes, the more rapidly the particles move, and the better are their chances of fusing together when they collide.

Since the temperature within a gas increases whenever the gas is compressed, we can see that if the star were to shrink, even by a relatively small amount, the temperature within the star would rise slightly. Even a slight rise in the temperature would produce a significant increase in the rate of nuclear fusion reactions, so the star would produce more kinetic energy per second at its center. The extra energy would tend to expand the star, until the center had cooled and regained its original size and rate of energy production. Likewise, if the temperature in the star's center were to decrease, the rate of nuclear fusion would also decrease, and the star would contract its core slightly. But this contraction would raise the temperature, increase the rate of nuclear fusion, and produce extra kinetic energy that would restore the original size of the star. This balance between self-gravitation and energy release allows stars to last for billions of years as natural "controlled thermonuclear fusion reactors," because each star's self-gravitation determines the star's central temperature and therefore its rate of energy production through nuclear fusion.

STARS' LIVES ARE RULED BY MASS

Stars lead lives of nuclear fusion, which turns energy of mass into energy of motion that spreads throughout the star, heating the star throughout its volume and producing visible light at the star's surface. Nuclear fusion makes visible, radiant energy from what was merely energy of mass. This works well for a few million, or even a few billion years, until the star exhausts its supply of nuclei to fuse together and therefore can produce no more energy of motion through nuclear fusion.

What distinguishes one star from another is the star's *mass,* the amount of material that the protostar managed to acquire when it began to contract to form a star. The amount of mass in a star determines the strength of the star's "self-gravitation," the gravitational force exerted on each small bit of the star by all the other parts of the star. More massive stars produce more self-gravitation, simply because they have more mass. In more massive stars, the greater self-gravitation "squeezes" the stars more effectively, and makes the particles within the star—mostly electrons, protons, and helium nuclei—move more rapidly in random directions.

The temperature of a group of particles measures the average energy of motion per particle: High temperatures mean that the particles dance randomly to and fro more rapidly, whereas a lower temperature means the velocities of the dancing particles will be more modest. The greater squeezing within more massive stars produce a higher temperature, with significant results.

In order for nuclear fusion to occur within a star, the nuclei must overcome their mutual repulsion. Although gravity pulls the nuclei together, their repulsion arises from the fact that all nuclei carry a positive electric charge, and it is a characteristic of nature that electric charges of the same sign repel each other. Every bit of additional speed helps in this effort. The rate at which nuclear fusion proceeds shows a tremendous sensitivity to the temperature. At "low" temperatures of less than about 15 million degrees Fahrenheit, nuclear fusion can barely occur: Almost no nuclei are moving rapidly enough to fuse together despite their mutual repulsion. At higher temperatures, where nuclear fusion does occur, every million degrees makes an enormous difference.

Our sun, a rather typical star, has a temperature at its center of about 27 million degrees Fahrenheit. This temperature, which arises from the interplay of the sun's self-gravitation and its rate of nuclear fusion, ranks slightly above the stellar average, just as our sun's mass ranks somewhat above the average for all stars. Stars with greater masses than the sun's have much greater rates of nuclear fusion, and therefore produce far more energy of motion from energy of mass during each second of their lives than the sun does. These high-mass stars are the luminosity lords of the universe, the stellar beacons that outshine all others. Bright stars such as Rigel, in the foot of Orion the Hunter, or Deneb, in the tail of Cygnus the Swan, have masses equal to ten to twenty times the sun's mass. They generate thousands of times more kinetic energy per second than the sun does. As the result of their enormous energy output, we see these high-mass stars as some of the brightest stars of the night skies.

Among the most profligate of the energy burners, we must include the star that exploded as Supernova 1987A, Sk −69° 202. Number 202 began its nuclear-fusing life with a mass about twenty times the mass of our sun. Astronomers can calculate that a star with twenty times the sun's mass will have a temperature at its center just over twice the temperature at the center of the sun: 60 million degrees Fahrenheit instead of 27 million. But because of the tremendous sensitivity of nuclear fusion to temperature, a star with a central temperature of 60 million degrees fuses nuclei not twice as rapidly, but *100,000* times more rapidly than the sun does. Since each nuclear fusion reaction turns the same amount of energy of mass into energy of motion, a star with twenty times the sun's mass produces 100,000 times more kinetic energy per second than the sun does. This energy of motion heats the entire star, so a twenty-solar-mass star therefore glows with a luminosity 100,000 times the sun's, and therefore stands out as a glorious beacon amidst a sea of lesser lights.

Stars pay a price for such glory. The most straightforward rule of the cosmos comes into play: There is no free lunch, and if you want to get energy of motion, you must get it from somewhere. If you get it from energy of mass, you must eventually run out of energy of mass. If you live fast, you die young. All stars begin to shine by fusing hydrogen nuclei (protons) into helium nuclei, and they can do so only so long as they possess a supply of protons. A twenty-solar-mass star begins life with twenty times more protons than the sun did. If the star fused protons into helium nuclei at a rate twenty times the sun's, it could last just as long as the sun. But if it fuses protons at 100,000 times the sun's rate (and it does!), then its twenty-times-greater supply of protons will hardly compensate. The twenty-solar-mass star will last only one five-thousandth as long as the star (twenty times the proton supply divided by 100,000 times the rate at which it consumes protons). How long is that?

Astronomers can calculate that the sun's supply of protons provides the sun with a total nuclear-fusing life of about 10 billion years. This result emerges from simple algebra: They know how much energy *each* fusion of protons into helium nuclei provides; they know how much energy the sun radiates each second as a result of such fusion; they can therefore calculate that approximately 4×10^{38} (four followed by thirty-eight zeros) protons fuse together each second in the sun's center. If the sun contained "only" 4×10^{38} protons—an enormous number, roughly equal to the number of protons in all the human beings on Earth—it would be good for only a second of nuclear fusion at its present rate!

But astronomers also know that the sun began with about 3 million trillion times more protons than a paltry 4×10^{38}! Astronomers have found the sun's mass (1.99×10^{33} grams) from observations of the Earth's speed in orbit around the sun, which depends on the sun's gravitational force on the Earth and therefore on the sun's mass: A greater mass would produce a greater speed in our yearly orbit. Most of the sun's mass consists of protons, and since astronomers know how much mass each proton has, they can calculate that the sun contains about 10^{57} protons. Astronomers also can calculate that about 15 percent of the protons reside, or will reside, in the sun's nuclear-fusing core. Astronomers can therefore see the future plain: After about 3×10^{17} seconds, the sun will have no more protons to fuse into helium nuclei. Now 3×10^{17} seconds equals 10 billion years. Since the sun and Earth are about 4.6 billion years old, we find ourselves nearly halfway through the sun's lifetime as a hydrogen-to-helium fusing star. A star with one five-thousandth of the sun's lifetime will last, not billions of years, but only about 5 *million* years before it exhausts its supply of protons. Then the trouble begins.

WHEN PROTONS RUN OUT

Every star that shines began its nuclear fusion by contracting to the point that its central temperature rises to at least 15 million degrees. The fusion of hydrogen to helium nuclei releases energy of motion and halts the contraction. Once nuclear fusion has begun, the star can maintain a steady rate of nuclear fusion, a constant luminosity, and a constant size, so long as it has an adequate supply of protons. It does this by balancing its tendency to collapse under its own gravitation against the energy of motion released each second by the nuclear fusion in its interior. But what happens as the star runs out of protons? Can it find other nuclear-fusion reactions to sustain itself? If not, what fate lies in store? These are the problems that lead to supernova explosions.

Every star has a problem as it exhausts its supply of protons. More precisely, the problem arises once most of the protons in the star's *core,* where nuclear fusion occurs, have been fused into helium nuclei. In a star like the sun, the nuclear-fusing core occupies no more than 1 percent of the star's volume. However, the *density* of the matter in the core far exceeds the average density, because the star's self-gravitation compresses the core most effectively. The sun has an average density of 1.4 grams per cubic centimeter—just 40 percent denser than water. But the *core* of the sun has a density of 150 grams per cubic centimeter, far denser than lead, mercury, gold, or uranium on Earth! Nevertheless, the core is entirely gaseous, not solid—a tribute to what a temperature of 27 million degrees Fahrenheit can do to keep particles in high-speed motion.

During a star's prime of life, the nuclear fusion turns protons into helium nuclei in the star's core. Once the innermost part of the star has been converted almost entirely into helium nuclei, the nuclear-fusing region begins to feed on the immediately adjacent material, which has grown denser and hotter. As a result, hydrogen-to-helium fusion will occur in a spherical shell around the nearly pure-helium center. Nuclear fusion will then spread slowly outward. Eventually, regions never before involved in nuclear fusion will participate in producing energy of motion. According to astronomers' calculations, the sun's nuclear-fusing core now contains about 15 percent of the sun's mass. As the sun ages, the core will shrink in *size,* but it will include a greater percentage of the sun's mass, until it embraces between 10 and 20 percent of the sun's total mass. This means that when the inner 10 or 20 percent of the sun has become helium nuclei instead of protons, the sun's days of easy living will be over.

RED GIANT STARS

As a star ages, its core shrinks to a smaller size, compressed by the star's self-gravitation, and grows steadily denser. The star's core contracts because of the dwindling supply of protons. To produce the same amount of energy from a diminishing supply of nuclear fuel, the star contracts and heats its remaining protons still further, so that nuclear fusion proceeds more rapidly than before. In fact, nuclear fusion occurs so much more rapidly that the star actually generates *more* energy each second than it did when it had a greater supply of protons. Driven by the laws of physics, the star behaves like an unwise motorist running out of fuel, who, in a hurry to reach the next gas station, floors the accelerator and consumes the remaining fuel supply more rapidly.

The star's increased rate of nuclear fusion produces more energy of motion per second within the star. This increase has two significant effects. First and most noticeably, the luminosity increases: The star grows a good deal brighter. Secondly, as the energy released in the core fights its way to the star's surface, spread among the nuclei by trillions of collisions, the flow of extra energy loosens and expands the star's outer layers. As a result, these outer layers grow larger and cooler. The expansion of the gas cools them to a lower temperature, just as squeezing the gas, billions of years before, made it hotter. At temperatures of a few thousand degrees, the star's surface glows mostly in red light, rather than the yellow or blue that characterized the star when its surface was hotter.

Thus the star becomes a "red giant," an aging caricature of its former self, with a bloated, reddened surface, cooled by its expansion to enormous size—a surface that conceals a shrinking, hotter interior. Antares, which forms the heart of the constellation Scorpius, the Scorpion, has grown to a hundred times the sun's diameter. Antares is now so large that if it replaced the sun, it would engulf the Earth itself, though Antares's outer layers are so rarefied that we would feel little resistance in our passage through our distended parent.

WHEN THE SUN BECOMES A RED GIANT

We would, however, feel the heat: Antares has 10,000 times the sun's luminosity (energy of motion released per second), and could melt the Earth. Our own sun will not become so fantastic a red giant as Antares. Five billion years from now, when our sun becomes a red giant as it begins to run out of protons in its core, the sun's luminosity will increase "only" a few hundred times, and its diameter will increase by a mere twenty to forty times, reaching perhaps to the orbit of Mercury. Even this will suffice to do "us" in, if "we" do nothing to combat the extra heat from the sun—heat that will increase the Earth's average surface temperature to about 1,000 degrees Fahrenheit. The quotation marks around "we" and "us" represent skepticism that human beings will persist in anything like their present form during the next 5 billion years. Nevertheless, if "we" are here, "we" shall have the chance to exercise our ingenuity to assure our survival, perhaps by arranging to move the Earth outward from the red giant sun. At ten times our present distance from the sun, near Saturn's orbit, we would receive about the same energy each second from the red giant sun that we do now from the prime-of-life sun of which we have grown justly fond.

THE PLANETARY NEBULA PHASE

If we move out, we must prepare to move back. The sun's red giant phase will last for a billion years or so, far less than the 10 billion years of its prime-of-life phase. During this billion years, the sun will maintain a high luminosity, and the extra energy of motion generated near its center will slowly push the red giant's outer layers still farther outward, as (for a time) the sun's tendency to expand wins out over its tendency to contract. The sun's outermost layers will expand past all the planets, past the billions of comets that orbit the sun, past the neighboring stars, until eventually the former outer parts of the sun, completely separated from it, will mingle with the rest of the diffused gas that permeates interstellar space.

During the relatively early phases of this evaporation, the sun will become a "planetary nebula," a star surrounded by a slowly expanding shell of gas (Figure 15). These nebulae have nothing to do with planets; astronomers of bygone eras misnamed stars with gas shells because, seen with the smaller telescopes of years past, each shell of gas looked something like the disk of a planet. In 5 billion years, the sun's outer layers will be pushed away from the sun's center—the result of the extra energy produced in the sun's central regions. Finally the sun's core will stand revealed, stripped of the millions of miles of gas that once shielded the core and received its energy, but which no longer form part of the star.

During its long red giant phase, and during the subsequent planetary nebula phase, which may last for millions of years, a star can lose a significant portion of its original mass. The star puffs that mass into space as it swells its outer layers to gigantic proportions. For a star like the sun, this mass loss amounts to no more than a few percent at most of the total, and therefore amounts to a mere footnote in the star's life story. In contrast, the stars that began life with many times the sun's mass lose a much larger fraction of their masses during their later, bloated existences. The star that became Supernova 1987A, for example, apparently lost about one-fifth of its original mass—a mass equal to four times the sun's mass—when it pushed its outermost layers into interstellar space. The star that remained, however, still had about sixteen times the sun's mass, and had the rest of its life to play out, until its final moment of glory ended its career as a star.

SUPERNOVA 1987A: EXPLAINING THE BLUE STAR THAT BLEW

In the spring of 1987, as soon as astronomers had identified the supernova as the former star Sk −69° 202, they faced a difficult task. Astronomers expected that (except for the Type I, white dwarf supernovae) an exploding star could arise only from a red giant that had exhausted all its fuel for nuclear fusion as it aged further and further. Hence they thought that a pre-supernova star—a star on the verge of explosion—must be a red giant, and "202" was clearly a blue, not a red star.

Theoretically minded astronomers returned to their computers. Soon they had confirmed, through calculations of how energy passes through a star, that some aging stars do not become red giants. Instead, because of relatively subtle differences in the types of particles that some stars contain (which affects how easily starlight can pass outward through them), some high-mass stars remain relatively compact and hot—therefore, blue rather than red—as they age.

Although both red and blue giant stars can explode as super-

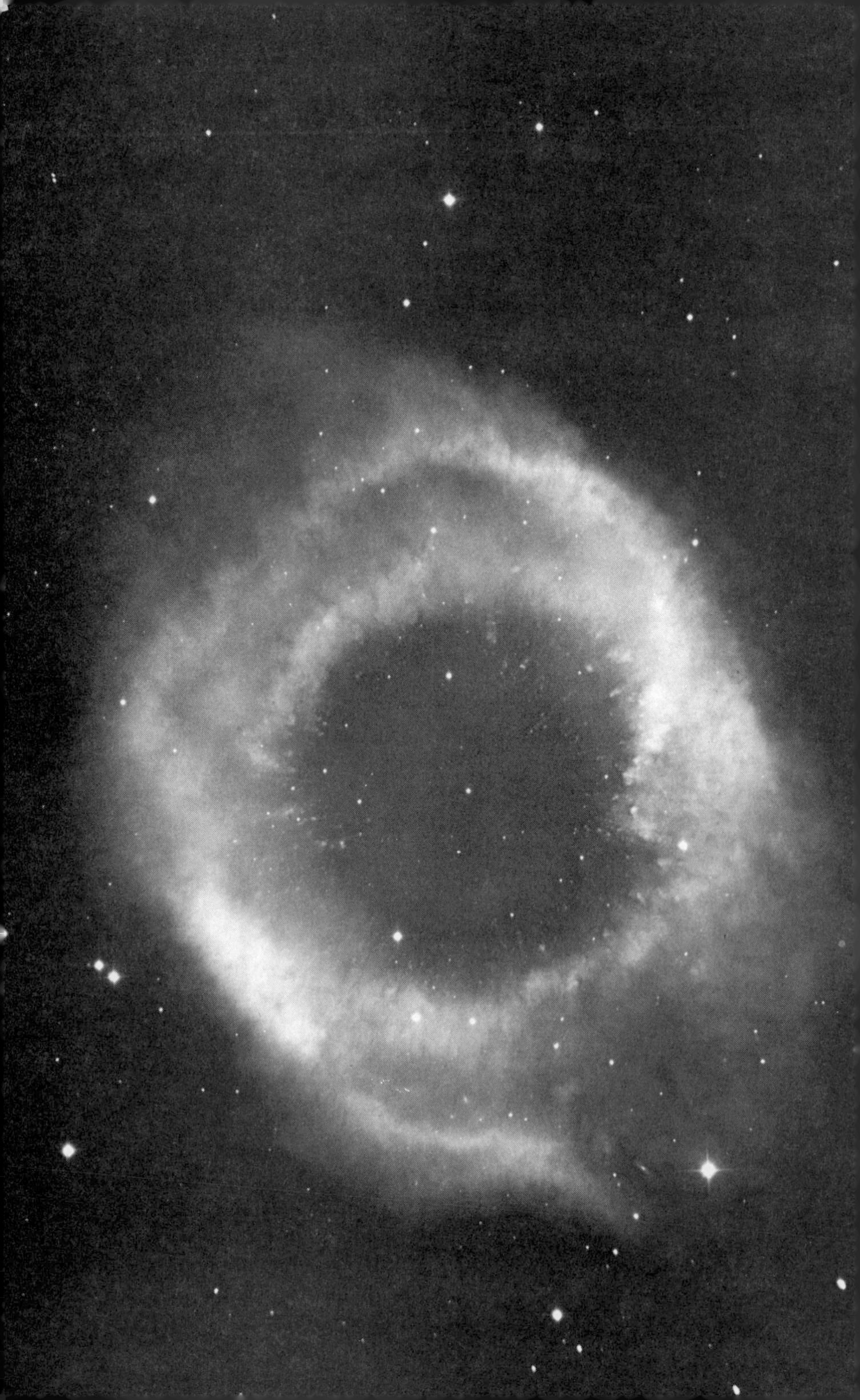

novae, a significant difference exists between an explosion in a red giant and in a more compact blue star. In the smaller, denser blue star, much more of the energy of the explosion goes into blasting the star's outer layers into space, rather than into producing the visible light that makes the supernova shine. Until the star's outer layers have expanded to a much larger size, they effectively block the light produced in a supernova outburst. A red giant's outer layers are already much more rarefied and farther from the star's center. Hence the blast can much more easily blow these layers into space, and more of the energy released in the explosion produces light. As a general rule, we therefore expect that a supernova explosion in a blue star will not shine with a luminosity as large as an explosion in a red giant.

SN 1987A was apparently the first Type II supernova seen on Earth that definitely exploded from a blue, not a red star. This provides an explanation of why SN 1987A never grew quite so bright as astronomers predicted when it was first discovered. The non-astronomer may wonder why it took SN 1987A to make astronomers perform the calculations to show the difference between a supernova explosion in a blue star and in a red giant. The reason is that astronomers work on what they know most about, and rarely venture into completely uncharted waters. Since they had not attained a full understanding of supernova explosions in red giants, they may perhaps be excused for not working much on the theory of how blue stars might explode, until one actually *did* explode. Furthermore, these calculations are quite difficult, even with the (absolutely essential) help of

---

FIGURE 15. This "planetary nebula" in the constellation Aquarius consists of the outer layers of an evolved star, pushed into space by the increasing energy output from the star itself. Because most of the star's output consists of ultraviolet radiation, the star, seen at the center of the nebula, appears quite dim in visible light. However, all of the light from the nebula arises because ultraviolet photons from the star have excited atoms within the surrounding shells of gas; these atoms then emit visible light as their electrons jump into smaller orbits. (Palomar Observatory photograph)

large computers. In fact, even before SN 1987A, astronomers knew from their calculations that some high-mass stars tend to remain blue as they age. If we seek to justify the ways of theoretical astronomers, we ought to take a closer look at some of them. A representative group can be found at the University of California at Santa Cruz, centered around a man with the odd name of Stanford E. Woosley.

7

THE THEORISTS

STAN WOOSLEY, born, raised, and trained in Texas, has lived for the past dozen years in Santa Cruz, California, making models of how stars explode. Woosley makes his models not in a workshop but on the computer, but with the same goal he had when he performed his early experiments in the shed behind his parents' apartment in Fort Worth: to see how things really blow up.

Woosley's boyhood laboratory work was so successful that the shed in question is no longer available as a national monument: His learning process burned it down. But from this and similar work Woosley evolved into a respectable astronomical theoretician. Today Stan Woosley is full-grown, a taller, more athletic version of Mick Jagger (who was born the year before Woosley in Dartford, England), with a head of curly light-brown hair (now suitably trimmed to professorial length) and a spring in his step that invites his students and colleagues to keep up with him as he walks through the magnificent, redwood-crowded campus of the University of California at Santa Cruz (Figure 16).

Santa Cruz, though one of the smallest and most intimate campuses in the vast UC system, is a campus conceived as a place for scholars to find peaceful meditation and rational intercourse, and the site of one of the most active astronomy programs in the United States. This rather remarkable state of events arose largely because soon after the campus opened in 1965 its astonomers began operating the university's Lick Observatory, located on Mount Hamilton some sixty miles away. Far from being one of the smallest academic departments, as is the case for astronomy on most campuses, the astronomy program at Santa Cruz ranks among the largest on the campus, a situation that has led to an entertaining irony: Some of the astronomers at Santa Cruz took the job not so much for the special ambience of the campus, but instead for the prominence of its astronomy program.

As a result of this anomaly, you will meet astronomers at Santa Cruz who seem unaware that they inhabit perhaps the most glorious acreage ever made into a college campus. A curious architectural fact compounds this clash of expectation and reality. At most universities, the astronomy department occupies the top floors of a science building, on the principle that astronomers should be close to the stars, an idea that is often reinforced by

FIGURE 16. Stanford Woosley of the University of California at Santa Cruz is one of the leaders in the effort to model star deaths on computers. (Photograph by Donald Goldsmith)

the practical fact that one or more telescopes for student use crown the building. But at Santa Cruz, the astronomers' offices fill the ground floor of one of the Natural Sciences buildings, hidden from the outdoors by a shadowy, collonaded walkway—as if the university had decided that this well-established rule likewise deserved testing.

Stan Woosley has an office on the third floor, because he is considered a physicist as well as an astronomer. Woosley spent his first twenty-nine years in Texas, first in Texarkana, later in Fort Worth, then in Houston for both undergraduate and graduate work at Rice University. With timing that seems breathtaking in retrospect, he arrived at Rice in the fall of 1962, as the United States' space effort swung into high gear for the race to the moon. President Kennedy visited the university and gave a special talk to the freshman class, urging them to participate in this race, and announcing that a "manned space flight center" would be built near Houston. Woosley, whose greatest fun to that date had come from testing which mixtures of potassium chlorate would produce the most effective explosion, decided to go into astrophysics.

By the time Woosley entered graduate school, Rice University had assembled an outstanding group of astrophysicists, drawn by the proximity of what became the Johnson Space Flight Center and by the generous working conditions at the university. Among this group were David Arnett and Donald Clayton, two experts in stellar evolution (the study of how stars age and die). Clayton was a Caltech product, a member of the famous group of theorists and experimenters led by Willy Fowler, now the grand old man of what happens in aged stars. Arnett had been a student of Alastair Cameron, then at Yeshiva University and now at Harvard, likewise an expert on stellar evolution, but with an emphasis on the early parts of stars' lives, as they condense from clumps of interstellar material.

In Arnett and Clayton, Woosley had instructors whose scientific skills and pedigrees could not be surpassed. Not surprisingly, after feeling cramped while writing a master's thesis on atomic transitions that involved a hundred pages of algebra, Woosley jumped at Clayton's suggestion that he study stellar evolution. (He might have jumped less quickly if Clayton had predicted that Woosley would eventually deal with computer programs that made a hundred pages of algebra look trivial.) A few years later, Woosley had written a thesis dealing with the processes that occur inside aging stars, and obtained his Ph.D. He became a postdoctoral fellow at Caltech for more than two years—"the best time of my life," he says—and went on to become assistant professor, associate professor, and then full professor at U.C. Santa Cruz. When the supernova's explosion reached the Earth, on February 23, 1987, Woosley was the department chairman, a post from which he soon obtained a leave of absence.

COMPUTERIZED MODELS OF STELLAR STRUCTURE

To an extent little appreciated by the public at large (how that complaint echoes through the world of science!), our knowledge of the universe relies on computer models, imaginary objects whose appearance, composition, and behavior with time are recorded within a large computer, the essential research tool of modern science. Capable of billions of calculations each second, such a computer can follow the evolution of an imaginary object—a model star, for example—to see what happens to the star, and how the star should appear to an observer thousands of light-years away. If actual stars correspond to the predictions of the models, then the models look good; if not, another generation of scientists (and graduate students) must continue their investigations of the universe by computer.

For the past three decades, ever since astronomers first laid hands on high-speed computers, they have used these machines to make increasingly accurate "models" of the interiors of stars. Such computer models use as their basic parameters the laws of physics—for example, Newton's law of gravitation—and the data that we have collected during the past fifty years concerning the fusion of atomic nuclei. With a high degree of accuracy, we now know what will occur at a given temperature and density when nuclei of any particular type have the chance to fuse together. These nuclear fusion data provide the cornerstone of any stellar model, since the essence of a star resides in its nuclear fusing core, from which newly released kinetic energy struggles outward through collisions among the particles in the star.

Among the primary aims of astrophysicists in making computer models of stars has been the desire to see whether the computer can make stars explode. More formally, astrophysicists sought to test the hypothesis, first put forward by Geoffrey and Margaret Burbidge, William Fowler, and Fred Hoyle, that a supernova explosion occurs when the core of a massive star fuses the nuclei up to and including iron, and then collapses. The "B^2FH" hypothesis (so called from its proponents' last names) has now passed its thirtieth birthday. Fowler, for his lifetime of work on nuclear fusion, has received the Nobel Prize. Although the computer models have vastly improved over three decades, they remain imperfect; but most astronomers agree that the B^2FH hypothesis appears to have been vindicated as the basic mechanism to explain supernovae.

Woosley is an expert at making computer models of stars on the brink of collapse, and his expertise has been honed by fifteen years of close collaboration with Tom Weaver, a physicist employed by the Lawrence Livermore Laboratory in California who knows as much about physics, high-speed computers, and how they can calculate mathematical models of stars as anyone around. The general plan of this research is straightforward. First, you write down the equations that govern what makes a star behave as it does.

- One of these equations keeps track of the amount of mass within each region of the star by noting the amount of volume, and the density of the gas (mass per unit volume) within that region.
- Another equation relates the change in gas pressure within the star to the amount of gravitational force at any particular point.
- A third equation describes how much energy is released each second as particles collide and fuse together within the star's core.
- And a final equation describes how easily the energy released as heat and light in this nuclear fusion can diffuse outward through the bulk of the star.

Once you have the equations, you must enter the information that specifies what a particular star is like—the star's mass, its size, and the abundances of the various types of particles inside it. In theory, once you have specified these quantities, the equations, calculated in a gloriously short time on the computer, will tell you the entire structure of the hypothetical star—which includes the density, temperature, pressure, and rate of nuclear fusion—at every point, working outward from the star's center, where nuclei fuse together, to the star's surface. Furthermore, the computer will track what happens as the star evolves, fusing hydrogen nuclei (protons) into helium nuclei at its center. This is the chief goal of stellar theorists: to understand the evolution of a star, that is, what happens to a star of a particular mass and coposition at every point in its life.

For all stars age and die. The 5-billion-year-old sun that we see now differs from the sun at birth, when the center of the sun became hot enough—about 27 million degrees Fahrenheit—for nuclei to stick together ("fuse") when they collided at high velocities. By now, so many of the hydrogen nuclei in the sun's central core have fused into helium nuclei that the sun has run through just about half its total lifetime as a steady producer of light and heat. Five billion years in the future, the sun will have exhausted the supply of helium nuclei in its core, and will begin to bloat and redden in its outer layers, even as its core shrinks and grows still hotter. This will make the sun a "red giant star," whose reddish, rarefied outer layers conceal an energy crisis developing within the star's central regions.

All this Stan Woosley knows in tremendous detail, but that is the easy part. The hard part is to figure out the stages of a star's evolution that follow the red giant phase. A host of questions crowd upon astronomers who want to follow a model aging star toward its death throes. Once all of the hydrogen nuclei in the star's center have fused to form helium nuclei, will the helium nuclei themselves fuse together to form more complex nuclei such as carbon? *Yes.* Will such carbon nuclei themselves fuse together to make still more complex nuclei? *Not in most stars, but in the most massive stars this fusion does occur.* Do the fusion processes within the star produce nuclei as complex as iron nuclei, and, if so, do these iron nuclei appear only in the center of the star? *Yes, in the most massive stars.* What is the distribution of nuclei in the rest of the star? *Like a multilayered onion, the star has its most complex and most massive nuclei, iron, at its center, and progressively less complex and less massive ones—sulfur, silicon, magnesium, neon, oxygen, nitrogen, carbon—distributed outward from the core.*

And then there arise still more spectacular questions and answers. Once a star's core becomes mostly made of iron nuclei—which simply cannot release *any* more light and heat by fusing together—what happens to the star's core? *It collapses under its own gravitational force in about one second's time.* What does the collapse produce? *Typically, a tremendously dense and compacted object called a neutron star, no more than a few dozen miles across.* What happens to the neutron star as the star's outer layers fall upon it? *It is first squeezed by them and then, like an overtight squash ball, it "bounces" to a somewhat larger size. This "bounce" triggers an exploding shock wave that, together with the neutrinos made during the collapse, blasts the outermost layers of the star into space.*

Which types of nuclei are made during the collapse? Which types are destroyed? How much of each? Which types of nuclei are made and destroyed by the explosive shock wave? How many X rays and gamma rays (high-energy forms of electromagnetic waves) are made during the explosive process, when the star expels its outer layers? How do these high-energy photons diffuse outward, producing the visible light from the explosion?

These and a hundred other questions have tentative answers from astronomers, but the search for the details remains a consuming one. The basic approach is straightforward. Take the computer models made by Woosley and others—computer pictures of stars at the ends of their nuclear-fusing capability—and make them explode. Ask the computer to calculate the characteristics of the explosion. For example, ask the computer to produce the explosion's "light curve," the hour-by-hour record of the amount of light produced by the exploding star, which will tell us how the star's brightness should appear to change as time passes. Also, ask the computer to calculate the star's "spectrum," that is, to determine which colors of light have been blocked by particular elements within the explosion that remove those particular colors.

Once you have the computer model of the exploding star, compare its characteristics with what you observe in an actual supernova. If the two match, the model must be good: Theory has another triumph. If they don't match perfectly (and in the case of supernova modeling they never do), return to the computer, and try again to obtain the details of how stars age to the point where they explode. Then follow that computer-created explosion to check the hypothetical star's characteristics against nature's own stars; that is, check the theoretical light curve and spectrum against the observed light curve and the observed spectrum of the supernova explosion.

Along with other astronomers, Woosley has had a plan like this in mind for fifteen or twenty years. Since astronomers can observe supernovae in distant galaxies, he was able to draw to some extent on data from about 600 supernovae that had been observed by modern astronomers. But because these supernovae were all relatively distant, the evidence from them lacked important clarifying details. Typically, for example, such supernovae were not observed until several days after their initial outburst, simply because they were not noticed until then. Thus astronomers lacked observations of the initial hours and days of the outburst, which could include data vital to the task of discriminating among competing theoretical models of exploding stars.

THE INTRUSION OF FACT

On February 24, 1987, Woosley came to work at his Santa Cruz office and was met by Phil Pinto, a graduate student whose thesis Woosley was then supervising. (Pinto is now a postdoctoral fellow at the Harvard–Smithsonian Center for Astrophysics.) Pinto showed Woosley the telegram that Brian Marsden had sent to several hundred astronomical centers, stating that a supernova had been detected in the Large Magellanic Cloud, magnitude 4.5 (about the same apparent brightness as the dim stars visible with the unaided eye), expected to reach magnitude zero (equal to the brightest stars of the night sky).

Woosley's initial reaction was not atypical. "That can't be right," he thought, "They'll have to cancel the telegram." (False alarms are not unknown to astronomy.) But when the supernova was still there on the following day, "it dawned on me that I should do something," said Woosley.

Woosley reacted as any red-blooded scientist would, and much as John Bahcall did, 3,000 miles away in Princeton, New Jersey. Woosley said to Pinto, "We've got to write a paper *right now*!" And Pinto, like any graduate student eager to succeed, could not but agree. Within two days, Pinto and Woosley had produced what became the first scientific monograph on the supernova, correctly identifying the star that had exploded, and pointing out that the short time interval between the detection of the neutrinos and the detection of light proved that the exploding star could not be a bloated red giant but instead must be a relatively small star, blue in color. A red giant star would be so large that the shock wave produced by the collapse of the star's core would take days to reach the star's outer layers, but in a smaller star, the explosion could reach the outside within hours.

In hindsight, Woosley has said that the Large Magellanic Cloud proved to be the best possible place (from an astronomical viewpoint) for a supernova to occur. A supernova in our own Milky Way galaxy would, of course, be closer, but its distance would be difficult to establish within a factor of two, because it is extremely difficult to determine the distance to an individual object in the Milky Way with higher accuracy. In contrast, from studies of hundreds of stars within the Large Magellanic Cloud, we know the distance to this galaxy to an accuracy of better than 10 percent. We can therefore translate the supernova's observed brightness into a luminosity—amount of energy produced per second—with a corresponding accuracy.

Furthermore, we observe the Large Magellanic Cloud with little obscuring effect from the interstellar dust in our own galaxy, which absorbs starlight with high efficiency. This dust concentrates toward the middle of the Milky Way, the central plane that bisects the galaxy's disklike shape. Because the stars in the Milky Way likewise concentrate toward this plane, a supernova in our own galaxy would be most likely to occur at or near the central plane, just where the obscuration becomes greatest. The supernova that produced the famous supernova remnant Cassiopeia A apparently exploded during the late seventeenth century, but the obscuration by interstellar dust was so great almost no one saw the supernova here on Earth; instead, we deduce the explosion by detecting radio waves from the remnant of the supernova, waves that easily penetrate interstellar dust.

THE EFFECT OF SUPERNOVA EXPLOSIONS ON THEORISTS' LIVES

On February 23, 1988, I joined Woosley, Pinto, and two UC Santa Cruz graduate students, Lisa Ensman and Dieter Hartmann, for a modest celebration (Szechwan food and Chinese beer) in honor of the supernova's first birthday. As is common in the small world of astrophysics, the participants' geographical origins spanned a wide range: Pinto, his thesis on supernova models nearing completion, is a New Yorker; Ensman, a hard-working, quiet student, comes from Indiana; and Hartmann was born close to the famous German university town of Goettingen.

For Pinto, the exploding star had been a nearly unalloyed blessing, inspiring him in his work and helping to advance his career. Of course, he had had to discard one thesis project, and to begin work on another of much greater immediate interest—Supernova 1987A. Woosley, already well set up at Santa Cruz, was aware that in a sense he had lost a year of his life, so excited that he barely had time to relax by windsurfing, let alone work on the stained-glass windows he loves to make. In addition, there was his not-so-beloved service as department chairman, which would again devolve on him now that his colleagues perceived that the supernova excitement was over. Ensman confessed that from her point of view, the supernova had burst on the scene a bit too soon—she was still taking graduate courses, even while attempting to master the computer program that would eventually be involved in her thesis research. Hartmann, a dark-bearded bear of a man, was the least directly affected: His research deals with sudden outbursts of "gamma rays," high-energy cousins of light rays, and he aims to detect and to analyze them; that is, he is more observer than theoretician. Since the detector used in his experiments was far from complete, Hartmann had not participated directly in the flush of supernova activity and emotion. But on that day Hartmann was in the best mood of all: He had just learned that NASA would fund the experiment for gamma-ray detection, so that in three years (all proceeding smoothly), he would be in business, an expert on "gamma-ray bursts."

Everyone loves an exciting new astronomical event, for it shows that the cosmos has once again obeyed the Scientist's Prayer: Now surprise me! To an established astronomer like Woosley, a supernova offers the chance to test the results of his career against physical reality, to improve his knowledge, perhaps to increase his power (never an unmixed blessing, for the responsibilities accompanying any such increase are unwelcome to a researcher), and to perceive fruitful new vistas of research that grow from the old ones.

To an advanced graduate student like Phil Pinto, the new supernova turned out to be just what the doctorate ordered: His years of training paid off in a few months, and although he would doubtless have become a respected academician in any case, the fortuitous timing of the explosion found him in the right place at the right time, his mind primed and agile, his computer programs fine-tuned by his thesis research under Woosley's supervision. It might have been otherwise. Had Pinto, for example, nearly finished a Ph.D. thesis demonstrating that only red giant stars are likely to produce supernovae, the appearance of Supernova 1987A from a blue precursor star would almost surely have called for a retracing of mental steps, new knowledge at the price of time lost over the old. As it is, however, Pinto's case stands as a prime example of fortune favoring the well-prepared, and his role as one of the masters of SN 1987A allowed him to become a supernova expert a few years sooner than otherwise predicted. To be sure, Pinto does not see things quite this way, for the supernova exploded before he was quite ready, and he had to revise many of the calculations he had performed, but from a more cosmic perspective, he was in the perfect spot at the right moment in life.

Lisa Ensman should also find SN 1987A a great gift of fortune, despite her temporary feeling of being swamped with so much to learn in so little time. Nothing drives forward the learning process more effectively than the need to play catch-up with others more highly qualified, and every graduate student in a theoretical science knows that he or she must lay down a foundation of well-established knowledge even while attempting to find a furrow of one's own to plow. Although SN 1987A is far from the exclusive theoretical property of the Woosley group at Santa Cruz, anyone within that group will have a head start at finding a thesis project based on the supernova, and should emerge not only with a Ph.D., but also with a claim to attention from the larger astronomical community.

Dieter Hartmann wil profit the most from SN 1987A in the long run. These are hard times for astronomical research in the United States. The glory days of the 1960s and early 1970s have long faded—witness the fact that the amazing series of planetary explorations made by the Pioneer, Viking, and Voyager spacecraft reached an end with the Neptune flyby in August 1989, when an automated vehicle launched a dozen years earlier visited the eighth planet, thirty times the Earth's distance from the sun. Not a single planetary exploration spacecraft has been launched by the United States between 1978 and 1988; the recovery from the Space Shuttle disaster left even the Magellan mission to Venus and the Galileo mission to Jupiter in a dangerous limbo. Although NASA now has ambitious projects underway to survey the cosmos with advanced satellite observatories, we may yet see the twenty-first century arrive with the Soviet Union and Japan in charge of Earth-based exploration of the cosmos.

Despite the recent difficulties in the United States' interplanetary research, the astronomical community reacted well to the challenge of Supernova 1987A. NASA managed to find several million dollars to support the increase in ground-based and satellite observations, and all astronomers agreed that this was not a time to say no to the chance to observe a once-in-a-lifetime (to put it mildly) occurrence. Dieter Hartmann will be well placed to pick up some of the fallout from this activity, as indeed he ought to be: We need well-trained researchers at the forefront of new technology and new research.

A brief survey and a lunchtime celebration hardly do justice to the thought and the work that lie behind our present understanding of supernova explosions. But when you read of what astronomers know, or think they know, about how stars grow old, and how and why some of them blow their outer layers into space, remember that this knowledge rests on the marvelous opportunity that society has provided some of its members, the chance to get ahead by understanding the universe.

8

STARS AT THE ENDS OF THEIR LIVES

SUPERNOVAE arise at the end of a star's lifetime—but only in certain types of stars, the ones born with especially large amounts of mass. (A minority of supernovae, the Type I supernovae, arise in white dwarf stars, which are discussed later.) In order to appreciate why a minority of stars explode, while the majority simply fade into ever-dimmer obscurity, we must consider the problems that a star faces as it ends its red giant phase. Most stars, our own sun included, find a cunning way to avoid total catastrophe. Other stars do not: They provide explosions, excitement, and the stuff of which life is made.

STELLAR EVOLUTION THROUGH THE RED GIANT PHASE

Consider a star that has twenty times the sun's mass and has begun to shine brightly. For examples of such a star, look up in the wintry sky at the three stars that form the belt of Orion the Hunter. Each of these stars is some 1,500 light-years from us, so we see them shining with light that left about the time when ancient Rome fell to the barbarian hordes. In their twenty solar masses, and in their future evolution, each of these three stars resembles the star that exploded in the Large Magellanic Cloud. Each of them has a luminosity—an energy output per second—nearly 100,000 times our sun's. If a star identical to the sun had the same distance from us as these three stars, we would need a high-class telescope to reveal the dim, sunlike star.

Stars with large masses have enormous luminosities. Their great masses squeeze their cores more effectively, producing higher temperatures and causing the nuclei to fuse far more rapidly than in low-mass stars. Most stars release energy of motion from energy of mass by fusing protons into helium nuclei, trillions upon trillions of times per second. Because of their high luminosities, the stars in Orion's belt will exhaust their nuclear fuel, the protons that originally formed most of the mass in the star, within a mere 10 million years after they began nuclear fusion. Their nuclear-fusing lifetimes therefore amount to only one-thousandth of the 10 billion years available to the sun, which exhausts its supply of protons much more slowly.

Since we remain ignorant of the birth date of the stars in Orion, we can only say that at some time within the next 10 million years, each of the three stars in Orion's belt will run out of protons in its core. As this happens, the star will begin to contract its core to fuse a dwindling supply of protons more rapidly, and will release even more energy per second than before. Some of the extra energy will enlarge the star's outer layers, cooling them as they embrace a larger volume, so that the star will become a cool, puffed-up red giant, similar to Betelgeuse, the red shoulder of Orion, or Antares, the brightest star in Scorpio, the Scorpion.

We have seen that during the first part of a star's red giant phase, all of the protons in the innermost part of the core fuse into helium nuclei. The core goes on contracting, and protons continue to fuse into helium nuclei in a spherical shell that surrounds the core. As the contraction proceeds further, more and more of the nuclei near the star's center fuse into helium nuclei. Eventually, the contraction makes the core so hot that the helium nuclei *themselves* fuse together. The fusion of helium produces carbon nuclei and some additional energy of motion.

BURNING THE ASHES

To fuse helium nuclei into carbon requires temperatures not of a mere 20 or 30 million degrees but of *300 million* degrees Fahrenheit. These higher temperatures are required because helium-4 *nuclei* repel one another (they each carry a positive electric charge) more strongly than protons do. Hence in order to overcome their repulsion and fuse together, the helium nuclei must be moving at higher velocities, which arise from higher temperatures. The continuing contraction of the star's core, produced by the star's self-gravitation, allows the star to raise its central temperature to 300 million degrees, and effectively to "discover" a new kind of fuel for nuclear fusion—the helium nuclei that were the ashes of its original fusion process.

Helium fusion represents an additional source of kinetic energy, but a rather poor one. Each fusion of helium nuclei into carbon nuclei releases only about 10 percent of the energy of motion that each hydrogen-to-helium fusion did. The star behaves like a camper who, after consuming good, dry logs in a campfire, puts the charcoal and ashes into a blast furnace and tries to obtain additional energy by burning them. They would indeed burn, but with relatively little result. Still, the star cannot be choosy: The fusion of helium nuclei into carbon both prevents the collapse of the star and also allows the star to keep on shining.

What happens when most of the nuclei in the star's core have fused into carbon nuclei? One might expect that the star would grow still hotter and begin to fuse carbon nuclei. However, in most stars, this does not occur. Instead, a strange and amazing process intervenes to bank the star's fires and to end its problems: The star becomes a white dwarf.

WHITE DWARFS

A white dwarf is what astronomers call the former core of a star that can no longer undergo nuclear fusion, but is kept from collapsing under its own gravity. Each white dwarf packs a mass approximately equal to the *sun's* into a volume about the size of the *Earth*. Thus matter in a white dwarf has a tremendous density, about a million times the density of water. A teaspoonful of white dwarf material, if brought to Earth, would weigh a ton! Thus a white dwarf testifies to the mighty powers of self-gravitation, for the matter became this dense because self-gravitation kept on contracting the star's core until the density rose to enormous values.

But what holds a white dwarf up? Why doesn't it collapse under its own gravity, since it releases no new energy of motion through nuclear fusion? A strange phenomenon of physics both halts nuclear fusion within the star, and also supports the star's core against its own gravitation. This phenomenon has the perky name of the "exclusion principle."

INVISIBLE STIFFENING BY THE EXCLUSION PRINCIPLE

The exclusion principle was the brainchild of Wolfgang Pauli, the same Austrian physicist who hypothesized the existence of the neutrino, a previously unobserved particle that could explain the observed effects of particle decays. Pauli's exclusion principle describes the fact that certain types of particles—most importantly for white dwarfs, electrons—simply cannot be packed any tighter once they have reached a certain critical density. This density is not the density at which the electrons are "shoulder to shoulder," stacked up one against the next. Instead, at a lower density than this, when the density reaches a mere million grams per cubic centimeter, the electrons will refuse any tighter packing. At that density, the exclusion principle provides a sudden stiffening, as if the particles went on strike to say, "This far and no farther will we be compressed." It does no good to attempt to squeeze the particles further: No additional amount of force, steadily applied, will suffice. You simply cannot force electrons into a smaller volume, once the density of matter reaches about 1 million grams per cubic centimeter, a million times the density of water.

Within the core of a star that becomes a white dwarf, the matter consists of carbon nuclei plus electrons. The carbon nuclei were made by nuclear fusion; the electrons formed part of the star when it formed and have not changed since then. Intriguingly enough, carbon nuclei by themselves are not affected by the exclusion principle, so if the core consisted only of carbon nuclei, it would go on contracting indefinitely. But the exclusion principle does affect the electrons, which cannot be packed into any volume that would cause their density to exceed the critical value forbidden by the exclusion principle.

If the electrons can't be packed into a smaller volume, neither can the carbon nuclei. Each electron has a negative electric charge, and each carbon nucleus has a positive charge. Since unlike charges attract one another, every electron attracts every carbon nucleus and vice versa. Once the density within the core rises to the point that electrons cannot be packed into a smaller volume, the electrons' attraction for the carbon nuclei holds the nuclei in place and thus refuses to allow the star's self-gravitation to squeeze them into a smaller amount of space. Therefore, once the exclusion principle becomes important, as the core contracts to a million times the density of water, it stiffens like cement that solidifies from runny liquid. In this case, however, the cement consists of carbon nuclei interspersed with electrons, packed to a density of a million grams per cubic centimeter and far stiffer than diamond.

Within the core nuclear fusion can no longer occur because the exclusion principle forbids it: The stellar "cement" keeps the nuclei from colliding with one another at the high velocities required for nuclear fusion. Deprived of nuclear fusion, the shrunken, dense core can produce no new energy of motion. And after a few million years, once the star's outer layers evaporate into space, the core stands revealed as a "white dwarf," the dying remnant of a once-active star, which slowly radiates into space the energy stored from its glory days of nuclear fusion. Millennium after millenium, the white dwarf becomes slowly but steadily fainter, but it will never contract further, for the exclusion principle can hold it up forever.

WHERE ARE THE WHITE DWARFS?

Our Milky Way galaxy probably contains about 50 billion white dwarfs. Astronomers base this statement on their belief that they understand how stars evolve, and on the fact that they have discovered several hundred white dwarfs, in orbit around some of the few thousand closest stars. From such breathtaking extrapolations are great astronomical conclusions made.

White dwarfs are born hot, since they were recently the cores of stars, and they gradually cool off, radiating the energy stored from their days as nuclear-fusing stars. As white dwarfs grow older and cooler, they continuously radiate less energy per second. They can therefore last a long time at the cooler stages of their existence.

The first-detected and most famous white dwarf accompanies the star Sirius, brightest of all stars in the night sky, the heart of Canis Major, the Big Dog who follows his master, Orion the

Hunter, through the skies of winter. Sirius is the fifth-closest star to the sun, a nuclear-fusing ball of gas with 2.3 times the sun's mass and twenty-three times the sun's luminosity. But among astronomers, what the public calls Sirius is designated "Sirius A," because the star has a white dwarf companion, "Sirius B," discovered more than a century ago, that shines dimly close to Sirius A.

From their observations of how Sirius A and Sirius B orbit around their common center of mass, astronomers have calculated that Sirius B has a mass almost equal to the sun's, a radius a bit larger than the Earth's, and a luminosity one-millionth that of Sirius A. Once Sirius B was a nuclear-fusing star much like Sirius A; today we require a powerful telescope to see the white dwarf, barely visible as it slowly releases its stored reserves of energy.

MASS LIMITS ON WHITE DWARFS

Not all stars become white dwarfs, because some stars have *too much mass* in their cores to achieve white dwarfdom. There is one additional strange fact about the exclusion principle, first discovered through independent calculations made by the Soviet astrophysicist Lev Landau and the Indian-born astrophysicist Subrahmanyan Chandrasekhar. The exclusion principle cannot support a star's core against its self-gravitation if the star's mass exceeds a certain amount of mass, approximately 1.4 times the mass of the sun. If a star has less than 1.4 solar masses left in its core after its red giant phase, well and good; the core can fade calmly away as a white dwarf. But if the core has more than 1.4 solar masses, called the Chandrasekhar mass limit after its discoverer, then the star is out of luck, for it simply cannot become a white dwarf. Such a star is primed for the supernova explosions we have been awaiting—the explosion that arises within a star that has run out of ways to support itself.

SUPERNOVAE FROM WHITE DWARFS

Before we turn to the heart of the supernova issue, white dwarfs must detain us a bit further, for it is now widely believed that some supernovae arise from white dwarfs themselves! A supernova involves the rapid release of an enormous amount of energy of motion. One way that such an energy release could occur involves nuclear fusion of material on an extremely rapid time scale within a white dwarf.

The theory of such white dwarf supernovae, the "Type I supernovae," runs as follows. Picture a situation in which a star becomes a white dwarf and then continues to have matter rain upon it. This can occur if the white dwarf is one member of a double-star system, a pair of stars born together and locked into an endless mutual dance, in which each star orbits the common center of the two stars' motions. Double stars are quite abundant; for example, Alpha Centauri, the closest star to the sun, actually consists of two sun-like stars, separated by a distance about equal to the distance from the sun to Saturn.

Suppose that in such a double-star system one of the two stars evolves into a red giant and swells its outer layers to enormous size. Some of the matter from this star's outer layers, attracted by the second star's gravitational force, will fall onto the second star. This infall of matter takes place whether the second star is a prime-of-life star or a white dwarf, but it has an important effect only in the *latter* case. Half a billion years from now, just this situation might arise in the Sirius system, where the white dwarf Sirius B orbits with Sirius A, which will some day become a red giant star.

In a double-star system where a red giant orbits with a white dwarf, the white dwarf's gravitation will attract some of the matter from the red giant's outer layers, which will accumulate on the white dwarf's surface. Because the white dwarf has such a small size and high density, the gravitational force at its surface is enormous—about a million times that at the surface of Earth. This huge gravitational force compresses any infalling matter to a high density and a correspondingly high temperature, just as compressing the air–fuel mixture in a diesel engine's cylinder heats the mixture until it ignites.

As the white dwarf's surface receives more and more matter, it will become hotter and hotter. Eventually the temperature will rise to the point that the protons fuse together. Then as the matter grows still hotter, helium nuclei will begin to fuse, producing carbon nuclei. Once the temperature reaches this level, both the infalling, new material and the old, white-dwarf material consist mainly of carbon nuclei. Nuclear fusion has made the new material into carbon, and the old material was already made mostly of carbon nuclei. At this point, all that has happened is that the white dwarf has become more massive as more material has been added to the object.

But as the object's mass grows toward the upper limit on any white dwarf, 1.4 times the mass of the sun, the white dwarf grows ripe for a sudden catastrophe. The exclusion principle generally acts (via the electrons) to hold nuclei in place and thus to prevent nuclear fusion. However, as more and more material joins the white dwarf, squeezing it more and more, the white dwarf grows denser and denser, hotter and hotter.

Finally, when the density reaches several *billion* times the density of water, and the temperature reaches nearly a *billion* degrees, the carbon nuclei suddenly fuse together, despite the exclusion principle. This sudden, violent fusion produces silicon nuclei, sulphur nuclei, and nickel nuclei: it also releases a tremendous amount of energy of motion in a fraction of a second, a sort of cosmic "carbon bomb," or, as astronomers call it, a Type I supernova.

Type I supernovae, those that arise from carbon bombs in white dwarfs, are observed to occur at about the same rate as Type II supernova. However, since Type I supernova are intrinsically somewhat more illuminous, they can be detected more easily than Type II explosions, so it seems likely that Type II supernovae in fact occur more often than Type I. Supernova 1987A was a Type II, not a Type I, supernova, so if we want to understand how and why it blew up, we must study what are probably the majority of exploding stars, the Type II, core-collasping supernovae.

MASSIVE STARS: NUCLEAR FUSION PAST CARBON

A star that has too much mass to become a white dwarf will explode at the end of its nuclear-fusing lifetime to produce a Type II supernova, a red giant whose core collapses. The key fact about the most massive stars is that they fuse nuclei *more complex than carbon.* They can do this because the matter in their cores is not dense enough for the exclusion principle to halt the fusion.

Why not? When astronomers make computer models of stars,

their models show an interesting wrinkle. High-mass stars have a density of matter at their centers *lower* than the less massive stars do. This is because high-mass stars, thanks to their much greater self-gravitation, can raise their central temperatures much more easily than low-mass stars. They therefore produce high temperatures in their core without having to produce so large a density of matter as that found in the centers of less massive stars.

A high-mass star will have a lower central density than a low-mass star not only during its prime of life, but also during its subsequent red-giant phase. The density in the core of a high-mass star, even after all the protons have all fused into helium nuclei, and after all the helium nuclei have fused into carbon nuclei, will not rise to a million times the density of water. Hence even at this stage the exclusion principle will not stiffen the core and prevent further nuclear fusion. The high-mass star will continue to contract its core, and nuclear fusion will proceed in the core, fusing carbon nuclei into still more complex nuclei, and producing a bit more energy of motion from these fusion reactions.

Thus in high-mass stars, in the same way that fusion in the core turns most of the helium nuclei into carbon, the carbon nuclei will eventually start to fuse with the remaining helium nuclei, and with each other. Here lies the secret of producing all the elements more complex than carbon, for these fusions produce nuclei of oxygen, neon, and magnesium. In addition, the fusion of carbon nuclei generates additional energy of motion, though carbon fusion generates less energy per fusion than the fusion of helium nuclei into carbon, and far less than the fusion of protons into helium nuclei that powers most stars. In an aging high-mass star, the newly released energy of motion supports the core against collapse, but only briefly—100,000 years or so after the helium nuclei have fused to form carbon. Then the nuclei, squeezed to a hotter and denser state, fuse still further, to produce nuclei such as aluminum, silicon, sulfur, chlorine, potassium, calcium, titanium, manganese, and finally iron.

THE END OF NUCLEAR FUSION IN MASSIVE STARS

Iron nuclei each have twenty-six protons and thirty neutrons, and so are called iron-56. Iron nuclei are not the heaviest or most complex nuclei, but they have a tremendous significance in the sequence of elements for a fundamental reason: They mark the end of the road in the production of energy through nuclear fusion.

Nuclear fusion "works" as an energy-generating process because the particles that emerge from the fusion have *less* mass, and therefore less energy of mass, than the particles that fused. Because energy can be neither created nor destroyed, the decrease in the star's energy of mass is balanced by the appearance of an exactly equal amount of new energy of motion that heats the star and makes it shine. This basic principle holds true for the fusion of all the nuclei lighter than iron: When they fuse, they reduce the star's energy of mass and increase its energy of motion.

But the world of nuclear fusion has an iron-clad rule: When you fuse small and simple nuclei together, you *release* energy of motion, and therefore gain extra energy of motion from the transaction; but when you fuse large, complex nuclei together, you sop up energy of motion, so you *lose* energy of motion in the deal. Iron nuclei mark the boundary in nuclear complexity between gaining or losing energy of motion through nuclear fusion. The fusion of nuclei *lighter* than iron reduces the energy of mass and increases the energy of motion. With iron, the rule is *reversed:* The fusion of iron nuclei *increases* the mass of the resultant nuclei, and therefore increases the energy of mass after the fusion, and it *decreases* the total energy of motion after the fusion process.

In short, the fusion of nuclei lighter than iron *releases* energy of motion, whereas the fusion of nuclei heavier than iron *consumes* energy of motion. Hence a star that has a core made of iron nuclei has run out of ways to release energy of motion, for it no longer possesses the sort of nuclei—nuclei lighter than iron—that turn energy of mass into energy of motion when they fuse together.

AN ONION RIPE FOR COLLAPSE

As a result of nuclear fusion, a massive star that has passed through its red giant phase, and whose innermost core has become mostly iron nuclei, has reached the moment of truth: It no longer can produce the energy of motion which, communicated to all the particles in the star through collisions among them, can support the star against its own gravitation.

We can picture the star's core at this point as a many-layered onion. The nuclei in the most central part of the core have become nearly all iron. Wrapped around this inner iron core is a layer of silicon nuclei and just outside that is a layer of mostly oxygen nuclei. Around the oxygen layer is a carbon layer, and around that is a layer where the nuclei are still mostly helium. Sprinkled in with the silicon nuclei are nuclei of sulfur, argon, and calcium, and among the oxygen nuclei are nuclei of neon and magnesium. All of these nuclei are interspersed with electrons, which do not participate in nuclear fusion. The entire stellar onion exists in this layered form because not quite all of the nuclei at its center have become iron nuclei through nuclear fusion.

By the time that the nuclei at its center become almost entirely iron nuclei, the star's core has shrunk to a size smaller than the Earth, with a diameter of only 3,000 or 4,000 miles. Finally, as almost all of the nuclei at the center fuse into iron nuclei, the core collapses under its own gravitational force, in only about *one-tenth of one second.* This brief time takes the exclusion principle by surprise. Before the exclusion principle can prevent the core from contracting past a certain density, critical events occur within the collapsing core—all on a time scale far less than a second.

THE BIGGER THEY ARE, THE HARDER THEY FALL

The collapse of a massive stellar core has several significant consequences. First, the collapse produces high-energy gamma rays that break the nuclei apart into individual protons and neutrons, the constituents of all atomic nuclei. Thus in less than a second, the collapse undoes the results of millions of years of nuclear fusion, which made larger nuclei from smaller ones, by stripping the nuclei back into its components, the protons and neutrons.

As the collapse proceeds still further, producing still larger densities of matter, the individual protons fuse with electrons to produce neutrons and neutrinos. At ordinary densities—even at the density within a white dwarf, a million times the density of water—you can't make electrons fuse with protons; but in the collapse of a star's core, the density rises to nearly *100 trillion* times the density of water, and the protons and electrons do fuse, making neutrons and neutrinos.

The neutrinos, which are massless, diffuse outward at the speed of light, leaving behind a stellar core now made entirely of neutrons. As described in Chapter 11, the entire core becomes a "neutron star," an object the size of downtown Chicago but with more mass than the sun contains, packed to a density 100 million times denser than a white dwarf! White dwarf matter has the density of a small mountain squeezed into an Olympic swimming pool. But to produce the density of matter in a neutron star, you would have to take the amazingly dense matter in such a swimming pool and then compress it to the size of your fingertip! Thus a cubic centimeter of neutron-core matter contains the mass of a mountain, and would weigh 100 million tons on Earth—if the Earth could support it. The exclusion principle does act on neutrons, and prevents the neutron star from contracting to less than a few miles in diameter.

NEUTRON STARS AND SUPERNOVA EXPLOSIONS

The matter that falls onto the newborn neutron star produces a supernova explosion. The innermost parts of the star's core collapse the most rapidly, and form a neutron star. Then, about one-fiftieth of a second later, the regions of the star surrounding the innermost core actually "bounce" off the neutron star and explode outward! This scenario has been investigated on high-speed, high-capacity computers; although the calculations do not seem to produce an outward explosion, most astronomers incline to the belief that the "bounce" does indeed occur. This works itself out in the following manner.

The core forms a neutron star because of the immense pressure that squeezes the collapsing star's inner regions. This collapse works a bit *too* well: The energy of motion of the infalling material squeezes the neutron star to a slightly smaller size than the exclusion principle will maintain once matter is no longer falling inward. Because of this, the neutron star, a fraction of a second after it forms, "bounces" to a somewhat larger radius, like a rubber ball that has been squeezed tightly and then released.

The outward bounce of the newly formed neutron star pushes

on the layers immediately surrounding it, and they in turn push on the layers surrounding them. Since the surrounding layers have progressively *lower* densities as we move outward, this push becomes progressively more effective. The push halts the original infall of the outer layers and reverses it into an outflow. Aiding the outward push are the neutrinos released by the collapse, which strive to move outward. Blocked by the dense matter, the neutrinos tend to push it to larger distances from the center. The outflow of matter proceeds at progressively larger velocities as we move outward in the star, because the same amount of energy from the bounce pushes on progressively smaller amounts of matter.

Computer calculations show that the bounce of the neutron core triggers a shock wave, a sudden increase in density and temperature, that moves outward from the core toward the star's surface. On Earth, shock waves meet us when an aircraft moves through the atmosphere at speeds greater than the speed of sound. The passage of such a supersonic aircraft produces a sudden rise in the density of the gas, and this shock wave surges past an observer on the ground to produce the clap of noise that we hear. In a star whose core has collapsed, the shock wave produced by the core's bounce carries much of the energy of motion released in the collapse outward through the star's outer layers.

The collapse of the star's core and the production of the intial outward-moving shock wave takes less than a tenth of a second. During that instant, the star's outer layers resemble the cartoon coyote who runs off a cliff and has time to think before he starts to fall. Because it would take many minutes for the outer layers to fall into the core—simply because these layers are far from the core—the core's collapse leaves the outer layers suspended in a frozen moment of time. These layers barely move inward during the few seconds before the shock wave hits them and blasts them outwards.

During the next few hours, while the shock wave makes its way through the star, it moves more and more rapidly, carrying the material it encounters along with it. As the shock wave nears the star's surface, its velocity reaches a significant fraction of the speed of light! The shock wave blasts perhaps the outer 30 to 50 percent of the star's mass outward into space. A tiny fraction of that mass—perhaps the outermost one-thousandth of one percent—explodes into interstellar space with velocities close to the speed of light. The result is a "Type II supernova," the common type of exploding star.

The shock wave that blasts the star's outer layers into space encounters nuclei that are mostly silicon, oxygen, carbon, and helium, with a sprinkling of nuclei such as aluminum, magnesium, nickel, and iron. But the explosion has such fury that during its initial violent moments, nuclear fusion occurs within the material being blasted outward. This nuclear fusion makes not only iron nuclei but even nuclei *heavier* than iron. To do this requires additional energy—the energy of motion of the explosion itself, some of which is transformed into the additional energy of mass of the heavier nuclei.

THE PRODUCTION OF "HEAVY" ELEMENTS BY SUPERNOVAE

Thus the pre-supernova star made elements up to and including iron, and the violence of the explosion itself made nuclei with *more* protons per nucleus than iron. Between the two ways to fuse heavier nuclei from lighter ones, supernova explosions have made essentially *all* the nuclei other than hydrogen and helium (though much of the iron stays behind in the collapsed core). We have seen that evolving stars can fuse nuclei up to iron, number 26 (twenty-six protons per atomic nucleus) in the list of elements that starts with hydrogen (one proton per nucleus) and helium (two protons per nucleus). If this were the full story, we would have no explanation of sixty-six naturally occurring additional elements—those whose nuclei contain anywhere from twenty-seven protons per nucleus (cobalt) through ninety-two protons per nucleus (uranium). The list of these sixty-six elements includes such important (to us!) entries as copper (number 29) silver (47), iridium (77), gold (79), mercury (80), and lead (82), as well as such rare elements as dysposium (66), ytterbium (70), and hafnium (72). We now know where the elements with a large number of protons came from: the first fraction of a second of a supernova explosion.

THE RARE ELEMENTS PAST IRON

When we look at the universe as a whole, we find that *all* of the high-number elements are extremely rare. Hydrogen and helium nuclei furnish at least 99 percent of all the mass that resides in atomic nuclei. The elements with three through twenty-six protons per nucleus, which include carbon, nitrogen, oxygen, silicon, magnesium, aluminum, titanium, chromium, and iron, have a total abundance less than one percent of the mass in hydrogen and helium nuclei. But the total abundance of all the *high-number* nuclei—all those with more protons per nucleus than iron's twenty-six—does not reach one *one-thousandth* of the mass of elements three through twenty-six. And most of the mass in the high-number nuclei consists of nickel (number 28), which is typically made in small quantities along with iron (number 26). Aside from nickel, the high-number elements have a total mass only one ten-thousandth of the mass of elements such as carbon, oxygen, silicon, and aluminum. When you look for these high-number elements—silver, mercury, gold, or uranium, for example—you are looking for the products of rare moments in the universe, the first moments after the explosion of a supernova.

WHERE DID THE EARTH'S ELEMENTS COME FROM?

When the Earth formed, close to our parent star, the warmth of the sun evaporated almost all the hydrogen and helium in our vicinity. As a result of the sun's warmth, the Earth today (except for water near its surface) contains almost *none* of the two most abundant elements in the universe. But the remaining elements were sufficiently heavy to avoid evaporation, and their relative abundances are much the same as we find in the stars: Oxygen, carbon, and nitrogen predominate; silicon, magnesium, aluminum, sulfur, calcium, and iron are somewhat less abundant; and everything heavier than iron and nickel appears only in trace amounts. Lead, which we don't think of as especially rare, has an abundance that falls short of iron's by a factor of half a million. Gold has only one-tenth the abundance of lead, and uranium has one-tenth the abundance of gold.

When we seek to mine any of the elements heavier than iron and nickel, we are searching for the remnants of isolated moments in cosmic history, the sudden shocks that began the explosion of supernovae. From these brief outbursts we must pry loose the elements we cherish for their special properties of luster, hardness, or radioactive decay. These properties arise from the nuclear structure of elements that were assembled during tiny fractions of a second, close to a newly collapsed neutron star, and then blasted into space through the same furious process that made them, a supernova explosion.

Some of these elements happened to occupy the regions of an interstellar cloud that later became the sun and its planets. And of the elements that made our planet, a tiny fraction have temporarily become part of our bodies. Every atom of calcium or oxygen, carbon, nitrogen, phosphorus, or iron—every atom except the hydrogen that forms part of our bodily fluids—has emerged from stellar furnaces, not gently but in nuclear frenzy, long *before* the Earth and sun formed, more than 4.5 billion years ago. These nuclei that we temporarily own take us back a long way in cosmic history. They not only connect us with the stars; they offer living proof that we would not be here without the stars that made us, along with all the other creatures on Earth.

COSMIC RAYS, MUTATIONS, AND THE EVOLUTION OF LIFE

Supernova explosions have done more than this. Life on Earth, and presumably life elsewhere in the universe, does not remain the same throughout time. Instead, it evolves new forms, and supernovae are involved. Evolution occurs because within a particular species different organisms have different rates of success in producing offspring. The struggle for reproductive success promotes the characteristics of the more successful organisms. As a result, the characteristics of the organisms successful at reproduction will eventually become those of the species as a whole, which we may call a new species if those characteristics differ sufficiently from the original ones.

But what makes one organism within a species different from others, sometimes significantly different? The answer is "mutations," random changes in the genetic makeup of individual organisms. Most mutations, far from being helpful to an organism's success at survival and reproduction, are downright harmful. These "mutants" quickly vanish from the scene of evolutionary struggle. But some mutations provide characteristics that help the organism to survive and to reproduce. If these are inheritable, the "mutant" will have many offspring, and so will the offspring have offspring, until, quite possibly, the "mutant" becomes representative of a new species.

And what causes mutations? No one knows for certain, but the answer seems likely to be "cosmic rays," misnamed during the 1920s, which are actually electrons, protons, and other nuclei that travel through space at nearly the speed of light. These fast-moving nuclei continuously bombard the Earth and the rest of the universe. Most of them pass through an organism when they encounter it, but it appears likely that some of the cosmic ray particles may sometimes strike the genetic material in an organism's "germ plasm" and alter it, thus producing a mutation. If this is so, cosmic-ray particles are a driving force behind evolution on Earth and elsewhere in the universe.

And what produces cosmic rays? The answer appears to be supernova explosions, although some particles may be accelerated to near-light velocities in interstellar space. The outermost layers of a supernova, blown into space at the highest velocities, may become cosmic-ray particles, traveling through interstellar space at nearly the speed of light. The bulk of the supernova explosion emerges at far more modest velocities, and eventually merges with other interstellar gas, enriching it in elements heavier than helium. But the fastest-moving particles speed on their random ways until they encounter something to stop them, perhaps an interstellar atom, perhaps a star, perhaps one of us. Thus the relationship of supernovae to the evolution of life on Earth appears to be straightforward. Supernovae make cosmic-ray particles; cosmic-ray particle impacts produce mutations; mutations drive evolution.

If this is so, supernovae do it all: They made our planet, they made our bodies, and they made the evolution that brought us here. We are living on the product, as the product, and by the product of stars that collapsed and then exploded, seeding the universe with their heavy elements and their fast-moving cosmic-ray particles. Far from being an isolated event, far distant from Earth and incapable of having any effect on us, Supernova 1987A can be seen as the latest in the chain of events that shaped our solar system, our Earth, ourselves.

LIFE ELSEWHERE

For those who find the prospect of a single example of life—life on Earth—stultifying in the long run, the explanation of how life arose and evolved on our planet offers some consolation. Nothing we have said here singles the Earth out as a particularly strange planet, subject to uncommon influences. Instead, astronomers think it likely (though far from proven) that many, perhaps most, stars have planets. Some of these planets are likely to be giant balls of gas, like the four giant planets of the solar system. Other planets, smaller and therefore more difficult to detect, may prove to be dense, rocky remnants that formed relatively close to their parent stars, whose warmth evaporated the closer planets' hydrogen and helium as they formed.

Cosmic rays certainly pervade our Milky Way galaxy, so far as we can tell, and would bombard nearly every planet that orbits one of the 400 billion stars in the Milky Way. The exceptions could be planets with particularly strong magnetic fields, which would deflect the cosmic-ray particles as they approach such a planet. (Since the particles would then slowly spiral in toward the planet's magnetic poles, this poses the delightful, though fanciful, prospect of a planet with a high rate of evolution near the magnetic poles and slow evolution elsewhere.)

If we imagine that the basic requirement for life, or at least the surest way to give life a chance, consists of a planet with a well-defined surface in orbit around a long-lived star, with a continuous bombardment of cosmic-ray particles to promote evolution, then we're in luck. The Milky Way may well have hundreds of billions of such sites. To verify this, all we need do is discover some rocky planets in orbit around some of the nearest stars. There too supernova-made elements may have formed themselves, under bombardment by supernova-accelerated particles, into living systems, capable of understanding how supernovae have changed the universe.

9

WHAT MAKES A SUPERNOVA SHINE?

ASTRONOMERS have a fairly good grasp—they believe!—of the basic mechanisms that make some stars collapse, and then explode their outer layers, when they exhaust all their ability to produce kinetic energy through nuclear fusion. One small detail remains to be explained: Why does a supernova shine in visible light for months after it explodes?

It may seem natural that when a star explodes its outer layers into space, that explosion should produce plenty of light. But reality does not unfold quite so simply. When a star's core collapses and then bounces outward, the energy from the explosion originates in nuclear fusion reactions, primarily in the reactions that fuse protons with electrons to produce neutrons and neutrinos during the collapse phase. More than 99 percent of all the explosion's kinetic energy—the energy contained in the motions of particles—appears in the energy of motion of *neutrinos,* elusive particles completely invisible to us. The explosion does produce a smaller amount of energy in the form of electromagnetic waves (photons), but most of this radiation originally consists not of visible-light, but of high-energy photons called gamma rays. What, then, explains the tremendous luminosity of a supernova in visible light—the luminosity that caused us to notice supernovae in the first place?

This question deserves serious attention, though we should recognize that, when we consider the total energy of motion released by the explosion, the energy in the supernova's visible-light output, which totals less than one percent of the energy of motion, *is* merely a detail. The question "What makes a supernova shine?" therefore amounts to asking, "Why does a fraction of a percent of the energy released by an exploding star appear in the form of visible light?" Astronomers think that they can answer this question, or can come close to the answer, but in the details of supernova light lies a mystery—a mystery that Supernova 1987A has helped to resolve.

The basic answer is that after the first few weeks, a supernova's visible light mainly arises from the radioactive decay of one particular type of atomic nucleus, cobalt-56. These cobalt-56 nuclei originate from high-energy collisions among particles during the supernova explosion. The partially resolved mystery is this: When cobalt-56 nuclei decay, they produce gamma rays, not visible light. What then makes a supernova shine in visible-light photons? In order to solve this mystery, and to understand the basic means by which exploding stars produce electromagnetic waves, we must examine the process by which supernovae produce the types of atomic nuclei that decay into other types.

PARTICLE COLLISIONS IN EXPLODING STARS

We have seen that the collapse of a star's core at the end of the star's nuclear-fusing lifetime overcompresses the core and produces an outward "bounce." This bounce triggers an expanding shock wave, a sudden increase in the velocity, density, and temperature of the gas that blasts through the star's outer layers, accelerating them to tremendous velocities. Within the gas that has been "shocked" by the passage of this blast wave, particles collide so violently that some of them fuse together as the star's outer layers are expelled into space.

When astrophysicists study particle collisions inside stars, they rely upon high-speed computers to deal with the immense number of collisions, each with several possible outcomes. But since trillions upon trillions of particles collide when a star collapses and then explodes, no computer can follow the enormous number of particle collisions one by one. Instead, the computer program reduces the model star to a much smaller number of particles, each one representing many trillions of real particles. It then models the totality of all the collisions with "Monte Carlo calculations." As if they were gamblers, the computers play the collision game over and over again. In each "game," the computer decides the outcome of each model collision by referring to a table of probabilities that provides the likelihood of each outcome, as determined by experiments in particle accelerators.

The computer sometimes chooses one outcome, sometimes another, in proportions set by the table of probabilities, so the entire model star collapses and explodes differently in the different games. But the astrophysicists then examine the different results to see which sorts occur the most often; these models are most likely to represent actual exploding stars. This "Monte Carlo method" cannot provide completely accurate results, but by playing the game a sufficient number of times, we can be reasonably sure that we know what will happen as the result of an enormous number of interacting collisions.

For our purposes, we can summarize the results of these computer models—which we believe provide us with answers about real stars—by identifying the chief types of nuclei that a supernova produces when it blasts its outer layers into space. The most important of these, by far, are three closely related elements: iron, nickel, and cobalt.

IRON, NICKEL, AND COBALT

On Earth, iron, nickel, and cobalt ores have a similar appearance and are often found in the same places, a boon to geologists who prospect for mining companies. But the association of these three elements goes much farther out in space and further back in time. Iron, nickel, and cobalt were made together inside exploding stars. Their intimate association—the way that one of these nuclear types turns into others—turns out to make supernovae shine.

We have seen that stellar cores collapse when nuclear fusion has transformed the nuclei within them into mostly iron nuclei. These iron nuclei are designated "iron-56": they each contain twenty-six protons and thirty neutrons. When stars have thousands or millions of years during which nuclear fusion can proceed in their cores, they indeed transform less complex nuclei into iron-56 nuclei. But when stars explode, and have less than a second to perform nuclear fusion before the explosion cools them off, the results are slightly different.

Most of the complex nuclei that emerge from nuclear fusion in a supernova are not iron-56 but a related nuclear species, nickel-56. Each nickel-56 nucleus has twenty-eight protons and twenty-eight neutrons. When exploding stars perform nuclear fusion on a time scale of one second or less, they are likely to make nuclei with *equal numbers of protons and neutrons,* because the star has nearly equal numbers of protons and neutrons to work with. The rapid nuclear-fusion processes in a supernova therefore tend to make nuclei such as carbon-12 (six protons and six neutrons), oxygen-16 (eight protons and eight neutrons), neon-20 (ten protons and ten neutrons), and silicon-28 (fourteen protons and fourteen neutrons), as well as large amounts of nickel-56. But the nickel nuclei differ from the others in this list in one important respect: They are unstable, and quickly decay into other nuclear types.

THE DECAY OF NICKEL NUCLEI

Nature has so arranged herself that nuclei come in two types, stable and unstable. The stable nuclei last forever (or at least far longer than the age of the universe), whereas the unstable types of nuclei "decay"—change their nature—after anywhere from a fraction of a second to many years' time. Nuclei such as carbon-12, oxygen-16, neon-20, and iron-56 are stable, but nuclei of nickel-56 are not. Instead, these nuclei decay about a week after they are formed into nuclei of cobalt-56. But each nucleus of cobalt-56 is also unstable, and decays after about two and a half months to form a nucleus of iron-56. The decay of nickel-56 into cobalt-56, and of cobalt-56 into iron-56, arises from what are called "weak forces" within each nucleus. These weak forces have the effect, in certain types of nuclei, of modifying the strong forces that hold nuclei together, so that the nucleus eventually changes one of its protons into a neutron, or one of its neutrons into a proton. Thus nickel turns into cobalt, which turns into iron. This process lights up a supernova.

If they could, a supernova would become marvelously bright in gamma rays, but invisible in what we call light rays.) But the gamma rays are trapped within the expanding shell of gas, because the gas blocks them. Lacking a clear path into the rest of the universe, each gamma ray collides with an atom that lies farther from the center of the explosion. In this collision, the gamma ray gives its energy of motion to the atom. This heats the atom, which collides with other atoms, so the gamma rays continue to heat the gas for months after the initial explosion. Because the expanding shell of gas remains at temperatures of thousands of degrees, it continues to emit visible-light photons, as any hot gas will, long after the initial heating from the shock wave has faded into a dim memory.

But the gamma rays that heat the gas arise from the decay of cobalt-56 nuclei. Thus, since fewer and fewer cobalt-56 nuclei remain to decay as time passes, fewer and fewer such decays occur, yielding a smaller amount of gamma-ray energy. Everything in this process therefore marches in lockstep with the seventy-seven-day half life of cobalt-56: We expect only half as much energy released from the decay of cobalt-56, and therefore half as much energy to heat the gas, each time another seventy-seven days have elapsed.

SUPERNOVA 1987A: THE THEORY VERIFIED

Before Supernova 1987A, this explanation seemed promising but not entirely proven. The measurement of supernova "light curves," the histories of the decline in the visible-light brightness of supernovae, had been made many times, but never with as fine an accuracy as astronomers sought. Nor could astronomers, before the space age, observe supernovae in ultraviolet, infrared, X-ray or gamma-ray emission.

The supernovae observed in our own galaxy (in 1604 or before) preceded the development of accurate techniques for measuring their brightnesses. But Supernova 1987A was close enough for its decline in light output to be followed for many half lives of cobalt-56—many times seventy-seven days. Supernova 1987A allowed astronomers to put the icing on the cake of the cobalt-decay theory of supernova light, and you will now look far and

THE HALF LIVES OF NUCLEI

If all the unstable nuclei in a supernova were made at one time (the moment of the explosion) and decayed at some later time, we might expect to see one or two brief flashes of light from a supernova, and nothing thereafter. But unstable nuclei do *not* all decay at the same time, so we see a supernova shining for months on end. The secret to nuclear decay lies in the concept of a nuclear "half life."

When unstable nuclei such as cobalt-56 decay, they do so independently of one another, on a probabilistic basis that prevents us from predicting precisely when any particular nucleus will decay. From experiments with unstable nuclei, physicists can say—and this is a key fact for supernovae—that cobalt-56 nuclei decay into iron-56 nuclei after seventy-seven days. However, this does not mean that if we have a thousand newly made nuclei of cobalt-56, they will *all* remain cobalt-56 for seventy-seven days and then turn into iron-56 nuclei. Instead, despite the fact that all cobalt-56 nuclei are essentially identical, each of them decays on a *random* basis. Some cobalt-56 nuclei take less than seventy-seven days to decay, some take exactly that time, and some take longer to decay.

Why, then, do we say that "a cobalt-56 nucleus decays after seventy-seven days"? The seventy-seven days refers to the nucleus's "half life," the time in which *half* of a *group* of cobalt-56 nuclei will decay. For any particular type of unstable nucleus, the half life—call it "T"—specifies the period within which *half* of the nuclei will decay. After a time T, the number of decaying nuclei will fall to one-half its original value; after another interval T, a further decline by a factor of two will occur, so the number of nuclei is only one-quarter the original amount. Another time interval T will produce another decrease by a factor of two, and so on, as the decays continue to occur randomly among the nuclei that have not yet decayed. This means after 3T of time have elapsed, the number of decaying nuclei will fall to one-eighth of its initial value; after 4T, the number equals only one-sixteenth of its original value; and so forth. Scientists call this process "exponential decay": Every additional time interval T halves the number in existence before that interval has elapsed. The rare type of carbon nuclei called carbon-14 provides an example.

Each carbon-14 nucleus has six protons and eight neutrons, and the half life of carbon-14 is 5,730 years. The decay of carbon-14 allows scientists to date organic material, such as old timber and old cloth, by measuring the ratio of the number of carbon-14 nuclei to the number of carbon-12 nuclei, which do not decay. Organic material assimilates carbon-14 nuclei from the atmosphere during the life of the organism. The decay of these carbon-14 nuclei after death can then reveal the time elapsed since death occurred.

COBALT-56: A HALF LIFE FOR SUPERNOVA EXPLOSIONS

As Supernova 1987A confirmed in detail, the brightness of a Type II supernova rises for several weeks to reach a peak, and then begins to decline. Within a few weeks after the peak, this decline shows the intriguing property called "exponential decay": The supernova's brightness falls by 50 percent during each seventy-seven-day period. When astronomers plot the supernova's "light curve"—the time history of the supernova's brightness—this decline in brightness appears as a *straight line* that slants downward to the right, because astronomers typically use the logarithm of the brightness in the light curve (see Figure 17). This straight-line decline in brightness is precisely what we would expect from the fact that cobalt-56 nuclei decay with a half life of seventy-seven days. Hence the detailed light curve of Supernova 1987A provided strong circumstantial evidence that the visible light from Type II supernovae originates from the decay of cobalt-56 nuclei.

LIGHT FROM COBALT DECAY IN SUPERNOVA EXPLOSIONS

To understand just why this proof convinced astronomers, we ought to ask how and why the energy from the decay of cobalt-56 nuclei appears as visible light.

When a star explodes, a shock wave blasts the star's outer layers outward, heating these layers enormously as it does so. This heating makes the layers radiate light and other electromagnetic waves, because hot gas does so naturally. If this were the entire story, a supernova would indeed shine, but after only a few weeks, the gas would cool so much from its expansion into a

much larger volume that it would cease to emit s amounts of electromagnetic waves.

But real supernovae are more complex than this. As wave heating dies away, the decay of cobalt-56 nuclei in nuclei becomes important. Each of these nuclear decays a high-energy photon called a gamma ray, and the en tained in the gamma rays keeps the supernova visible f on end.

Gamma rays from cobalt decay make a supernova cause the high-energy photons cannot escape directly i

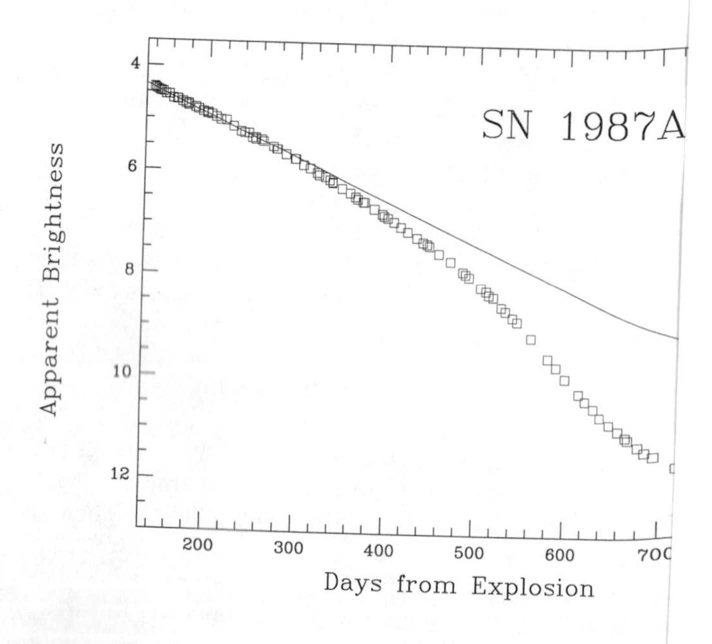

FIGURE 17. The "light curve" of Supernova 1987A recor nova's changing brightness as observed on Earth (using a system that measures the logarithm of the apparent brig inverse scale, so that lower numbers imply greater brightnes zontal time axis shows the number of days past the dete neutrino burst on the morning of February 23, 1987. (Cou ert Kirshner)

wide (though you may succeed eventually) to find an astronomer who does not consider it far more likely than not that supernovae owe their light—at least after peak output—to the decay of nuclei of cobalt-56.

From the moment that Supernova 1987A was discovered, astronomers strove to record its changing brightness accurately. Their measurements of apparent brightness included not only the supernova's visible-light output, but also the significant amounts of energy emitted as ultraviolet radiation. Astronomers aimed to measure what they call the supernova's "bolometric luminosity," the amount of energy produced in all the photon types that make up electromagnetic waves. Their task was far from easy.

Like other exploding stars, Supernova 1987A emitted photons of all types—gamma-ray, X-ray, ultraviolet, visible, infrared, and even some radio photons. However, most of the exploding star's energy output in photons consisted of visible light and ultraviolet. Hence in order to measure the supernova's "bolometric light curve," the history of the explosion's total energy output in electromagnetic waves of all types, astronomers needed the IUE satellite for ultraviolet observations as well as ground-based astronomical observatories for visible-light measurements. Within a few weeks, however, the supernova had cooled to the point that it no longer emitted large amounts of ultraviolet.

Figure 17 shows the light curve of Supernova 1987A—the result of nearly two years of observation of SN 1987A. Once the supernova reached its peak luminosity, about three months after it exploded, "its light curve began to search for its radioactive tail," as the astronomer Stan Woosley puts it. That is, the decline in the supernova's brightness was expected to reflect directly the decay of cobalt-56 nuclei, which occurs with a seventy-seven day half life. For this reason, astronomers expected the supernova's light curve to show a straight-line decline on their graphs, which plot the logarithm of the supernova's brightness against the time since the explosion. Indeed, before the end of June 1987, such straight-line behavior appeared, and astronomers could then calculate with good accuracy the amount of cobalt-56 involved in the radioactive decays that produced the energy that made the supernova shine. The observation that the supernova's light curve showed just such straight-line behavior, with a half life of seventy-seven days, confirmed with stunning success the hypothesis that the light in this Type II supernova—as in others—arises from the decay of cobalt-56 nuclei.

FALLING BELOW THE STRAIGHT LINE OF BRIGHTNESS

Close to a year after its detection, during the winter of 1988, the supernova revealed another phenomenon that had been forecast from astronomers' models of exploding stars. At this time, the light curve based on visible light and ultraviolet observations fell *below* the straight line predicted by a model based on exponential decay with a half life of seventy-seven days (Figure 18). This occurred for a simple reason. A year after the explosion, the expanding gas shell of gas had declined in density to the point that it could no longer block all the gamma rays produced by the decay of cobalt-56 nuclei. This trapping had occurred because the density of gas was high enough to trap all the gamma rays, but now a fair fraction of the gamma-ray photons could escape directly into space.

From that time onward, the supernova could no longer convert high-energy, gamma-ray photons into visible-light photons with 100-percent efficiency. When the gas was dense enough to "catch" all the gamma rays produced in cobalt decay before they could escape into space, all of the energy in the gamma rays went into heating the gas, which then radiated that energy in the form of visible light. But as some of the gamma rays began to escape without heating the gas, the light curve shown in Figure 18, which does not include the gamma-ray output from the supernova, began to show a decrease more rapid than a straight line decline. This result occurred as the supernova's gamma rays began to carry a significant portion of the supernova's output directly into space.

HOW MUCH COBALT DOES A SUPERNOVA NEED?

As we have seen, the perfect *match* between the half life for the decay of cobalt-56 nuclei, seventy-seven days, and the half life observed in SN 1987A's light curve (the time required for the supernova's brightness to decline by 50 percent), which also equals seventy-seven days, supports the conclusion that the light from the supernova arises from the decay of a host of cobalt-56 nuclei. To complete our understanding of the production of supernova light from cobalt-56, astronomers have calculated the answer to the question, "How *much* cobalt-56 did the supernova produce in its exploding outer layers?"

If you want to answer this question, you must first know the distance to the supernova (which we do), in order to find how

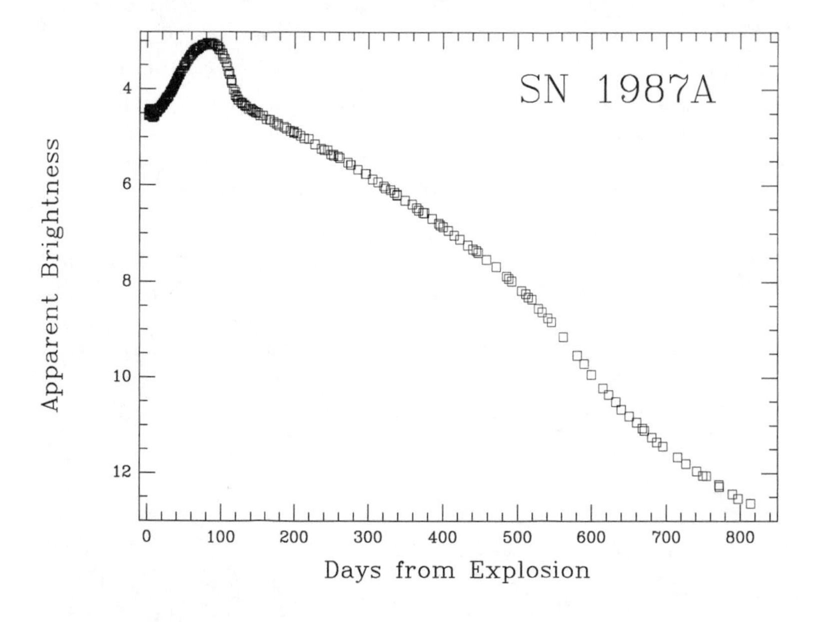

FIGURE 18. At about 300 days past the first detection of the supernova, the light curve shows a decline in brightness more rapid than the straight line that had previously been the rule (see Figure 17). This deviation from straight-line behavior almost certainly resulted from the fact that after 300 days, some of the gamma rays produced by the beta decay of cobalt-56 nuclei could escape directly into space instead of being trapped within the supernova's expanding shell of gas. (Courtesy of Robert Kirshner)

much energy the supernova was radiating each second to produce the brightness that we observe. A greater distance would imply a greater amount of energy per second, since without such greater energy, the object would appear fainter, simply because its energy output must spread through a greater volume before reaching us. We must also know (which we do) how much energy arises from each decay of a cobalt-56 nucleus.

Armed with knowledge of the supernova's distance, of its observed brightness, and of the amount of energy of motion released in each cobalt-56 decay, astrophysicists could calculate that the explosion of SN 1987A must have produced about 3×10^{54} (three followed by fifty-four zeros) nuclei of cobalt-56, if gamma rays from the decay of cobalt-56 are the source of the supernova's light after the first few weeks. To put this number of nuclei in perspective, we may note that the mass contained in 3×10^{54} nuclei of cobalt-56 totals 7 percent of the mass of the sun, or about 23,000 times the mass of the Earth! In other words, Supernova 1987A turned 23,000 Earths' worth of matter into cobalt-56 nuclei. Actually, the explosion first made the nuclei in question into nickel-56, and these nickel-56 nuclei decayed, after a few days' time, into the cobalt-56 nuclei whose continuing decay powers the supernova's light output. By now, when more than 99.9 percent of the cobalt-56 nuclei made in the explosion have decayed, the supernova appears considerably dimmer than it did in early 1987.

Our theoretical models of exploding stars suggest that the outer layers of SN 1987A had a total mass of fifteen to twenty times the mass in the sun. Therefore, if a mass equal to 7 percent of the sun's mass fused into cobalt-56 by the explosion, about half a percent of all the matter in the star's outer layers was transformed into cobalt-56 nuclei by nuclear fusion. Astronomers find this conclusion entirely reasonable. Theoreticians' hearts have swelled with pride over the fact that SN 1987A was expected to shine with the light from cobalt decay, and it did so, with an amount of cobalt entirely reasonable from our understanding of what makes stars explode.

DIRECT OBSERVATION OF COBALT IN SUPERNOVA 1987A

A final, direct confirmation of the cobalt-decay model of supernova light came when astronomers aboard the Kuiper Airborne Observatory (the "KAO") detected infrared waves from cobalt atoms. We know that nearly every type of molecule and atom—those in our bodies, for example—will emit infrared waves, long-wavelength cousins of visible light, at temperatures near room temperature. The exact wavelength and frequency of the infrared radiation provides a cosmic fingerprint, a tip-off to the type of molecules and atoms that produce such radiation.

During the fall of 1987 and the spring of 1988, astronomers aboard the KAO detected infrared waves from SN 1987A at various wavelengths and frequencies. Some of this infrared emission arises at precisely the wavelength and frequency known to be produced by cobalt atoms. The astronomers who interpret these observations now feel certain that this emission arises from the cobalt made in the supernova explosion. Some of the cobalt-56 nuclei have captured electrons to form atoms, and some of these atoms radiate infrared waves. Thus, by detecting cobalt atoms in the material exploded from the supernova, the KAO's infrared observations have furnished another piece of evidence in favor of the dominant theory of how supernovae produce their light: Nuclei of cobalt-56, decaying with a half life of seventy-seven days, do the job quite well. We have now seen not only the energy released by the radioactive decay of cobalt, but infrared waves produced by some of the cobalt nuclei that have managed to form atoms before they decay.

LIGHT ECHOES FROM SUPERNOVAE

An intriguing sidelight on SN 1987A appeared a year after its detection, in February of 1988. As had been predicted by some astronomers, a series of "light echoes" became visible around the supernova.

A light echo, in astronomers' parlance, arises when light rays from a particular source have taken a roundabout path to the observer. Since light rays travel in straight lines (we are not dealing here with the effects of gravitational forces on light), this roundabout path consists of two straight-line segments, but not in the same direction. Instead, photons from the explosion have traveled outward in all directions, and have met with interstellar dust grains.

Dust grains are ubiquitous in interstellar space. A typical interstellar grain contains a million or so atoms and spans about a hundred-thousandth of an inch. Such grains apparently form in the outer layers of stars during the stars' red giant phases, and are slowly expelled into space along with the rest of the stars' outer parts. Floating in interstellar space, dust grains can both scatter light—send it in a different direction at random without otherwise changing the light—and also absorb (completely swallow up) some of the light that strikes them.

The scattering of light by dust grains produces the light echo. At a particular time after the explosion, an observer will see light that was scattered from dust grains somewhat out of the direct line of sight, then scattered onto a path that leads directly toward

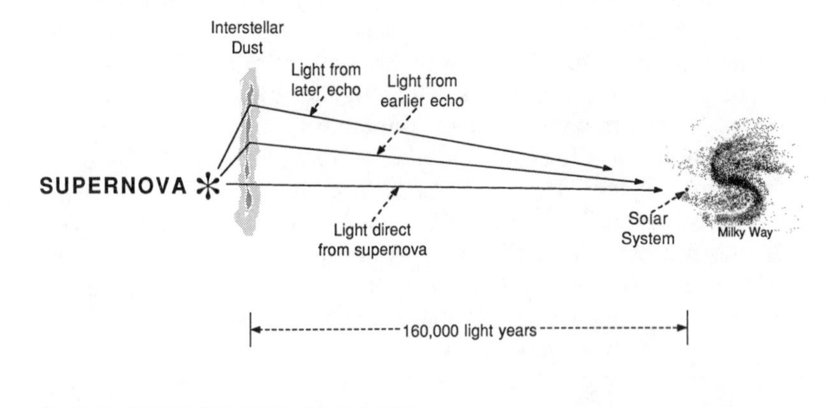

FIGURE 19. The light echo arises from the fact that dust in the general vicinity of the supernova scatters light from the explosion. At any given time, some light scattered by dust reaches us (or another observer). Since the scattered light has taken a longer journey on its way to us than the light traveling in a straight line, we see the "original" outburst, as scattered by the dust, in a delayed manner—hence the term "light echo." This diagram greatly exaggerates the deviation from straight lines of the light rays that create the light echo. (Drawing by Marjorie Baird Garlin)

the observer (Figure 19). By the time these photons reach the observer, they have taken somewhat longer (say, about a year longer on a journey of 160,000 years) than the photons that arrived directly, without scattering by dust grains.

As a result of this scattering, at a time one year past the explosion the observer will see those photons whose two-line path totals one light year longer than the direct path. Two years after the detection of the explosion, photons that have traveled two light-years more than the direct photons will arrive, and so forth. This will be true for all observers; it is not a question of being fortuitously placed so as to see a light echo. Rather, the observer sees the echo from just those photons that happen to have been sent in exactly the observer's direction by reflection from a dust grain.

Figure 20 shows a photograph (printed as a negative to bring out more detail) taken with the 3.6-meter reflecting telescope at the European Southern Observatory in Chile, using a state-of-the-art imaging system that reveals details never before observable. In this photograph, the bulk of the light from SN 1987A has produced the overexposed black area at the center. The two concentric circles are the light echoes, scattered light that has taken several months longer to reach us than the light from the central image.

Astronomers have been quick to make mathematical models of the situation that gives rise to the light echoes. Their calculations show that the inner ring arises from dust located about 400 light-years from SN 1987A, and the outer ring from dust grains some 1,000 light-years from the supernova, all still well within the Large Magellanic Cloud. The fact that we see rings of light implies that the dust causing the echoes does not have a uniform distribution. If the distribution were uniform, we would see no gap between the inner ring and the outer ring, but instead would see a relatively smooth sheet of light around the supernova. Eventually, astronomers hope to use the changes in the light-echo patterns to *map* the distribution of dust in the parts of the Large Magellanic Cloud.

The light echoes from SN 1987A provide an additional fillip, previously predicted but never before observed in a supernova, to add to the excitement surrounding the explosion. Astronomers

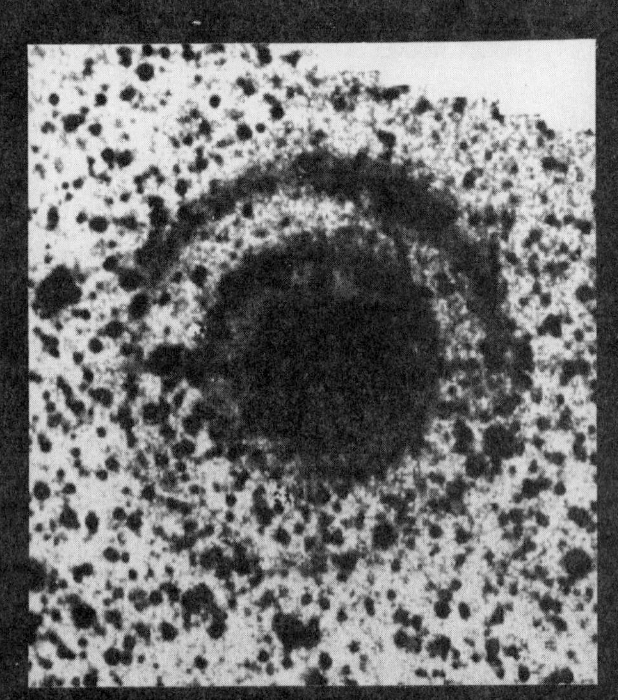

13 February 1988

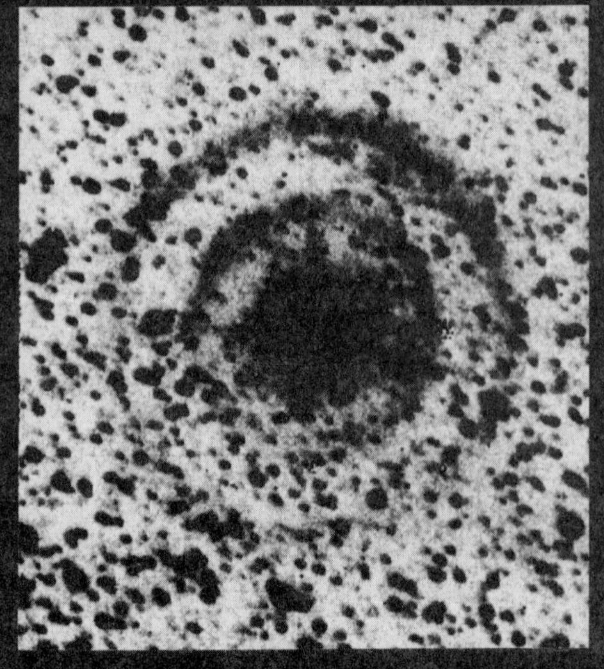

16 March 1988

find the light echoes intriguing but not particularly significant. They care far more about the processes that govern a supernova explosion, and the photons and nuclei that result from such explosions, than about the pale scatterings of a supernova's visible light. In short, they seek the substance not the shadow, and so may we.

FIGURE 20. The "light echo" from SN 1987A appears as two "shells" of light around the supernova. The appearance of these shells changes with time, as revealed in these photographs from February and March of 1988. (Photograph copyright by European Southern Observatory (ESO), reproduced by permission of ESO)

10

NEUTRON STARS, PULSARS, AND BLACK HOLES

SUPERNOVA explosions bring an end to a star's life as a true *star:* No longer will the fusion of one type of nucleus into another liberate energy. But though the star may have exhausted itself so far as nuclear fusion is concerned, from its ashes there may arise one of the strange beasts of the cosmos, either a neutron star or a black hole.

Before we plunge into the cramped confines of such objects, we should note that some supernovae, probably including most of the Type I (white dwarf) supernovae, succeed in obliterating themselves; that is, they explode so violently that they blow their cores completely apart, leaving behind nothing of consequence. However, according to the best theories of astrophysicists, nearly every collapse of a star's core leaves behind a remnant of immense density and tiny size. This remnant, at most a few dozen miles across, contains more mass than the *sun.* We are talking about a stellar cinder packed to enormous density, whose gravitational force may continue to influence its surroundings significantly.

Astronomers' calculations of stellar core collapses have yielded an assortment of useful models. Like much of scientific research, these models answer many problems but leave others unsettled. The question of which collapsing stellar cores will form black holes and which form neutron stars falls somewhere between the settled and unsettled categories. Astronomers think that they know the fate of stars fairly well in an approximate sense, although they are still at work filling in the details. In order to understand the latest theories concerning the collapse of stars, we must consider what these collapses can produce; that is, we must address the basic issues of what black holes and neutron stars may be.

BLACK HOLES AND GRAVITATIONAL FORCES

Black holes represent the last sink of stellar failure, the ultimate collapse, from which there is no return (at least, as astronomers hasten to add, in anything like the present lifetime of the universe). A black hole is an object with such an enormous gravitational force at its surface that nothing—not even light or other types of electromagnetic radiation—can escape.

Notice the three little words "at its surface" in the sentence above. As Isaac Newton showed (and here Albert Einstein agreed entirely), gravity always attracts, and the strength of the gravitational force exerted by any object decreases as the distance from the object's center increases. To be precise, the force of gravity exerted by an object falls off in proportion to the *square* of the distance from the object's center.

We on the Earth's surface, nearly 4,000 miles from the Earth's center, each feel a certain amount of gravitational force. If we were to double our distances from the center of the Earth, to a distance of 8,000 miles, the Earth's gravitational force on ourselves would diminish to one-quarter of its present value. At 100 times our present distance from the center, the force would decrease to one ten-thousandth of what we feel now. At such a distance, we might fairly say that the Earth's gravitational force on us would be unimportant, but we can recognize that the Earth, like every object in the universe, would continue to exert a gravitational force of attraction upon us, small though it may be.

Suppose, on the other hand, the Earth began to shrink (as unlikely as anything in this uncertain cosmos). Once our planet shrank to half its present radius, *losing no mass as it did so,* all of us on Earth's *surface* would find ourselves at half our present distance from the Earth's center. As a result, although neither the Earth's mass nor our mass had changed, we would each feel four times more gravitational force, so our weight would increase fourfold, a heavy burden to carry on our shrunken, denser Earth.

In contrast to our experience *on* the Earth, a satellite in orbit *around* the shrinking Earth would feel no change in the amount of gravitational force from the Earth, simply because no change in the distance from its center to the Earth's center would occur. Instead, the satellite would continue in the same orbit (now much higher above the "shrunken" Earth's surface), unaffected by the tremendous changes in gravitational force felt by those of us on the Earth's surface. The gravitational force between two objects depends only on three factors: the mass of the first object, the mass of the second object, and the distance between their centers. Greater masses imply greater gravitational forces, and greater distance, as we have seen, reduces the amount of gravitational force. Therefore, if we change neither the masses of the objects nor the distance between them, the gravitational force that one exerts on the other remains unchanged, even if one (or both) of the objects shrinks or expands.

In the real universe, the Earth will not collapse, but stars do, at least in their cores. Before such a collapse, the stellar core, with a mass greater than the sun's mass, may have a diameter of some 10,000 miles, roughly the same as the Earth's. The core's collapse well deserves the name, since calculations of how a dying star behaves show that the core contracts in diameter by a factor of a thousand, so that it shrinks to a diameter of only ten or twenty miles.

THE DENSITY OF MATTER

The density of any object—the measure of how tightly mass has been packed into the object—equals the object's mass divided by its volume. If the water in a cubic centimeter has a mass of one gram (and it does), the density of water equals a gram per cubic centimeter. Because space has three dimensions, the volume of any object varies in proportion to the *cube* of its diameter. Thus the density within a collapsing stellar core will increase in proportion to the *cube* of the factor by which the diameter decreases. If the diameter shrinks to one-thousandth of its original size, the volume occupied by the core falls to one-billionth of its original value, so the density within the core rises by one billion times! In the fraction of a second that sees the core collapse, the density rises from a "mere" billion times the density of water to densities so enormous that they rise completely beyond what we can grasp intuitively. A star's collapsed core, as we have seen, takes matter that once could have made a small mountain and squeezes it to the size of a fingertip. The universe is strewn (at enormous intervals) with these tiny but massive stellar fingertips.

The gravitational force exerted by the core on any object at its surface must rise enormously as the distance from the surface to the core's center declines drastically. After the collapse, *all* of the mass in the core lies within a few miles of the surface, whereas before the collapse, much of that mass was thousands of miles away from the center. A collapse by a factor of 1,000 in an object's size produces an increase of *1 million* in the gravitational force at the object's surface. In the twinkling of an eye, gravity rises in strength to the point that it can significantly affect light waves.

LIGHT, GRAVITY, AND EINSTEIN

For light too feels the effects of gravity. This fact, one of Einstein's great contributions to science, arises from the fact that gravity effectively bends space. In the presence of a massive object, space that would otherwise be "flat," so that light traveled in straight lines through it, becomes "curved," as if the object made a dimple in space, just as a heavy ball placed on a sheet of taut plastic would dimple the plastic. Light rays then travel through space, not along straight trajectories but on paths that bend toward the object, like a small ball rolling past the dimple on the sheet that changes its course because of that dimple. Einstein's theory of relativity showed that the closer the approach of the light to the massive object, the greater would be the amount of the bending of light by gravity.

Albert Einstein created this theory in Berlin during the years 1915 and 1916, as the First World War raged through Europe. Einstein's general theory of relativity, first published in 1916, received dramatic confirmation in May 1919, when a British expedition traveled to Africa to observe a total eclipse of the sun. A few months later came exciting news: Careful measurements of the positions of the stars whose rays passed close by the sun during the eclipse showed that the sun had indeed bent the trajectories of light! The stars' positions differed from those revealed on previous photographs, taken at times when the sun did not happen to lie in almost the same direction as those stars. Einstein was delighted; he wrote to his mother in Switzerland, "Good news today . . . the British expeditions have actually proved the light shift near the sun." When the world learned about the the eclipse results, Einstein awoke to find himself the world's most celebrated scientist, a mantle of fame he wore with dignity for the next thirty-six years.

BEYOND EINSTEIN: OPPENHEIMER

During the 1930s, as scientists became accustomed to Einstein's theory of general relativity, they speculated about what would happen to a star that shrank to a fantastically small size. In 1939, J. Robert Oppenheimer, who a few years later would direct the Manhattan atomic bomb project at Los Alamos, wrote a paper with Hartland Snyder, a graduate student at the University of California, Berkeley, entitled "On Continued Gravitational Contraction."

In this paper, Oppenheimer and Snyder analyzed the equations that Einstein had written to describe the effect of an object's gravitational force on the space around the object. They knew that light escaping from a gravitational field of force loses energy. Our intuition suggests that if light loses energy, it must slow down, but our intuition fails us: The light continues to travel at the same speed (186,000 miles per second). The loss of energy appears as a decrease in the *frequency* of the light waves, and as a corresponding increase in their *wavelength.* What if the gravitational field grows stronger? The light must lose still more energy, and its frequency decreases still further. Eventually, if the field is strong enough, what was visible light before its escape can escape only as infrared waves, with frequencies below all visible-light frequencies. And if still more gravitational force exists? Then the light must lose even more energy. Eventually, in a sufficiently strong gravitational field, the light must lose *all* its energy, and no light will escape at all.

Einstein's equations show how this loss of energy in a gravitational field is related to the bending of light when rays of light pass close to a massive object. Oppenheimer and Snyder took the next step, and showed that if a star collapses to a sufficiently small size, then no light, no other types of electromagnetic radiation, no other types of particles—nothing but the continuing gravitational force—can escape from the object. The term "black hole" was coined to refer to such a beast, an object with such immense gravitational force at its surface that nothing escapes. For an object with the sun's mass, calculations based on Einstein's theory show that the critical radius equals *two miles:* If the object has this radius or less, it will be a black hole.

Before the Second World War, Oppenheimer and Snyder's work looked like fantasy, and indeed when I was an undergraduate, "black holes" were regarded as simply the playthings of theoretical physicists. I well remember the first serious talk I heard about black holes, about 1965, when I was a beginning graduate student in Berkeley, delivered by the eminent Princeton University astrophysicist John Wheeler, the man who had named black holes. When Wheeler used this term, a graduate student next to me snickered, "of Calcutta?" Today the situation is reversed: Everyone knows the black holes of space, so few the historical background of British India.

DO BLACK HOLES EXIST?

There's a reason why they're not laughing now: Astronomers think that they have found several regions in space where a black hole is in orbit with a relatively normal star, possibly after a supernova explosion produced the black hole in a double-star system. Furthermore, in the core of our own Milky Way galaxy, and in several neighboring galaxies, evidence is mounting for the existence of "supermassive black holes," black holes with millions, or hundreds of millions, of solar masses. Such supermassive black holes may prove to be the gravitational "seeds" around which galaxies like our own have formed. Although the verdict remains uncertain, most astronomers would agree that black holes with starlike masses and diameters of only a few miles may prove quite common in galaxies such as our own.

NEUTRON STARS: NOT QUITE BLACK HOLES

Black holes, exciting though they may be, do not arise from the collapse of most stars' cores. Instead, according to the best calculations of astrophysicists, such a collapse usually produces a *neutron star,* a stable, immensely dense object made almost entirely of neutrons. Not quite so small as black holes—for, if they were, we could never observe them, and would know almost nothing about them—neutron stars nevertheless are starlike masses packed into city-sized regions of space.

What keeps this collapsed object from shrinking still further under the influence of its immense gravitation? Physicists have a double-barreled answer: Part of the support comes from "strong forces," the forces that act between nucleons (protons or neutrons), and part from the exclusion principle, which, as we discovered in Chapter 8, is the reason that white dwarfs don't collapse. The exclusion principle describes how neutrons resist being packed beyond a certain critical density, no matter how hard you may try to do so. Since neutrons have no electric charge, this resistance cannot arise from the repulsion of particles by electromagnetic forces. Indeed, the exclusion principle is not what physicists call a force at all; it provides no ability to repel particles when the density falls below the density at which the exclusion principle comes into play.

To imagine how the exclusion principle "works" in a neutron star, think of a giant, invisible hand with countless fingers that does nothing to the neutrons until the density of matter rises to about 10^{15} (one thousand trillion) grams per cubic centimeter. Once this occurs, the hand puts a finger on every neutron to hold it in place, and thus halts all further tendency of the star's core to contract. Simultaneously, strong forces acting between the neutrons also act to resist any further compression. The entire stellar core resembles a single giant atomic nucleus, but one made of about 10^{57} neutrons, a number so much larger than the number of dollars in the national debt that one grows tired simply by writing it out. As in an atomic nucleus, strong forces hold the particles in place, and the exclusion principle provides additional stiffening.

MASS LIMITS ON NEUTRON STARS

Strong forces and the exclusion principle give the neutron star a nearly eternal existence. But there's a catch: Objects that contain more than a certain mass cannot become neutron stars. (This is reminiscent of the mass limit on white dwarf stars that we discussed in Chapter 8). If you attempt to form a neutron core with an amount of matter greater than three to five times the mass in the sun, even strong forces and the exclusion principle won't protect you against collapse. Stated more accurately, the collapse of a star's core that contains more than three to five times the sun's mass can never produce a neutron star. Instead, since the exclusion principle cannot halt the collapse of such a massive object, a black hole will form. Here we have a clear indication (to be sure, from theoretical considerations alone) as to why some stellar collapses produce black holes, others neutron stars. The answer lies in the mass: High-mass cores collapse to form black holes, lower-mass cores yield neutron stars.

PULSARS FROM NEUTRON STARS

Neutron stars have a claim to fame in addition to their immense densities and tiny sizes. Rotating neutron stars emit flashes of electromagnetic waves, and are the objects responsible for "pulsars." Pulsars are sources of radio waves (and often of visible-light, ultraviolet, and X-ray emission as well) that *pulse* regularly. These objects do emit relatively low amounts of some electromagnetic waves at all times, but in addition, they produce "spikes" of much more intense radiation at regularly spaced intervals. During the late 1960s, as pulsars were discovered, their amazingly regular pulsations roused speculation that the long-awaited "little green men" had revealed themselves. Astronomers found it difficult to see how any naturally formed object could emit radiation pulses at exactly spaced intervals of time. But the answer soon became clear: Rotating neutron stars produce beams of electromagnetic radiation, like the rotating beam of a lighthouse. As these beams sweep by us, they produce the "pulses" of radiation that characterize pulsars.

CONTROL OVER SPIN: THE CONSERVATION OF ANGULAR MOMENTUM

The rotating neutron stars that produce pulsars are typically rotating several times per second—quite rapidly for an object the size of Manhattan! How did such rapid rotation arise? The answer lies in a familiar fact of daily life, which scientists call the "conservation of angular momentum." This principle states that an object free from outside forces will spin more rapidly if it contracts, less rapidly if it expands. If the object contracts in size by a factor of two, it will spin four times more rapidly; if it contracts to one-tenth its original size, the rotation rate will rise a hundredfold.

An ice skater absorbs this principle in his or her bones, knowing that contraction of the body to a smaller size will cause more rapid spin. Likewise, an acrobatic high diver (or an unsuccessful ski jumper) knows full well that even though gravity pulls the entire body downward, any spin superimposed on this motion follows the principle of angular-momentum conservation. Hence a contracted body spins more rapidly, and to slow the body's spin to achieve a correct entry into the water, the body must extend to full size.

Stars work the same way. Compress a spinning star and it will spin more rapidly. Most stars rotate, because they contracted from regions of an interstellar cloud of gas that had some tendency to rotate. But for mature, nuclear-fusing stars, this rotation has little effect. Our sun, for example, spins once a month, nothing to get excited about, but not zero either. But the conservation of angular momentum makes it clear that if the sun shrank from its present diameter of more than a million miles to a diameter of twenty miles—by a factor of 50,000 let us say—its rate of spin would increase by the square of 50,000, or 2.5 billion times! Instead of rotating once per month, the sun would then rotate 1,000 times per second, a considerable difference.

THE IMPORTANCE OF MAGNETIC FIELDS

The collapse of a star's core increases not only the star's rate of rotation but also the strength of its magnetic field. Just as contraction to one-tenth of its original size will make a star spin 100 times more rapidly, the contraction will likewise magnify any magnetic field at the star's surface by a factor of 100. But as we have seen, the collapse that yields a neutron star involves a contraction in the star's diameter by factors of many tens of thousands! Such a contraction increases the star's magnetic field by billions of times. The collapse therefore produces a neutron star that rotates many times per second and also possesses an immensely strong magnetic field, far stronger than anything we can generate on Earth.

A rapidly rotating neutron star, whose powerful magnetic field rotates along with it, has an immense effect on any charged particles—for example, protons and electrons—in its vicinity. Charged particles feel electromagnetic forces from the rotating magnetic field, and are swept into motion along with it. In fact, the magnetic field accelerates any charged particles close to the neutron star to nearly the speed of light. Once the particles reach this velocity, they emit waves of electromagnetic radiation because of the "synchrotron emission" process discussed below.

Relying on theoreticians' best work, astronomers conclude that some regions near the surface of a rotating neutron star have more charged particles, or a more intense magnetic field—or both—than the average. As the neutron star rotates, these regions are "hot spots"—sources of more intense emission than average. Each time the neutron star spins, the radiation from one or more such hot spots creates an intense pulse of electromagnetic waves that repeats itself almost exactly in every rotation.

SYNCHROTRON EMISSION: THE SIGN OF COSMIC VIOLENCE

What is this "synchrotron emission" that produces electromagnetic waves when neutron stars rotate? Synchrotron emission draws its name from a particular type of particle accelerators on Earth. These accelerators allow us to smash particles into one another at nearly the speed of light, and to study the results of these collisions in an attempt to understand the composition of matter at the smallest levels of size. Like a rotating neutron star, a synchrotron uses magnetic fields to accelerate electrically charged particles, flinging them around and around a circular track that may be many miles long.

During the 1950s, as synchrotrons set new records for the most energetic, fastest-moving particles on Earth, scientists realized that they had created a phenomenon never before observed on Earth: A strange glow of light emerged from the charged particles as they moved at nearly the speed of light. Theoretical physicists had not been idle, and furnished the explanation. Whenever a charged particle, such as an electron or a proton, moves through a magnetic field at nearly the speed of light, and changes either its speed or its direction of motion (or both), electromagnetic waves will appear. These waves of "synchrotron emission" draw their energy from the particles' rapid motion, so unless energy is continuously supplied to the particles moving at the speed of light, they will slow down from the act of producing synchrotron emission.

Physicists soon saw that the photons produced by synchrotron emission have a characteristic distribution in energy, that is, in the relative number of photons that are produced at different energies, frequencies, and wavelengths. Once again, the distribution of photon energies provides a cosmic "fingerprint," revealing what type of process has produced the photons. The distribution of energies in synchrotron-emission photons differs from the energy distribution of photons produced by "thermal emission," the radiation of electromagnetic waves by any hot object. Thermal-emission photons have an energy distribution that rises to a peak at a particular frequency and then falls almost to zero at higher photon energies. Synchrotron-emission photons, in contrast, have an energy distribution that simply declines smoothly at progressively higher photon energies.

Physicists could apply what they learned in the laboratory, and in their calculations, to the cosmos at large. Because of synchrotron-emission photons' characteristic distribution in energy, you can look at photons that have traveled thousands, or even millions, of light years, and can recognize that they have been produced by the synchrotron emission process. Here we have one of the great detective methods beloved by astronomers in their attempts to unravel the mysteries of the cosmos. For the photons produced by synchrotron emission are the mark of cosmic violence.

In order for the synchrotron process to produce photons, four requirements must be met: You must have charged particles, they must be moving at nearly the speed of light, they must move through magnetic fields, and they must change either their speed or their direction of motion. Three of these four requirements are satisfied nearly universally. The cosmos contains plenty of charged particles, and nearly everywhere in space we find some sort of magnetic field, though often a weak one. Particles often naturally change either their speed or their direction of motion, or both. But it *is* unusual to find particles moving at nearly the speed of light. This requires violence: the sudden release of energy.

During the 1950s, astronomers made detailed observations of the filaments of gas in the Crab Nebula, a prominent remnant of the supernova seen in explosion in the Milky Way in the year 1054 (see Figure 11 on page 73). When they observed the ratios of the numbers of photons produced at each different photon energy in the Crab Nebula, these ratios showed the mark of violence: Most of the light from the Crab Nebula's filaments arises from synchrotron emission. This discovery propelled the Crab Nebula into a prominence (astronomically speaking) that it has never lost.

By now, thirty years after astronomers realized that the Crab Nebula uses the synchrotron process to produce photons, astronomers routinely analyze the photons from a distant object such as a peculiar galaxy or a supernova remnant, and can conclude that the distribution in energy of those photons reveals that the synchrotron-emission process must be at work. Therefore an explosion must have accelerated particles to nearly the speed of light.

An additional useful detail of synchrotron emission deals with the different types of photons—radio, infrared, visible light, ultraviolet, X ray, and gamma ray. These photons differ in their energies, from the lowest energy per photon (radio) to the highest (gamma-ray photons). To produce *any* type of photons by synchrotron emission requires that charged particles be accelerated to *nearly* the speed of light, which means in practice to more than 99 percent of that speed, 186,000 miles per second. But if you accelerate particles to "only" 99 percent of the speed of light, you will get mostly radio photons from synchrotron emission, and precious few photons of visible light. In order to produce significant amounts of visible-light photons, you need speeds that are closer to 99.9 percent, or even 99.99 percent, of the speed of light. To reach these greater velocities requires that you put much more energy into the particles than is required to reach a "modest" 99 percent of the speed of light, because as you approach the speed of light, every fraction of a percent increase in velocity requires far more energy. Unsurprisingly, then, most of the objects that emit photons through synchrotron emission turn out to be emitting mainly radio photons and few others: Their violence is relatively mild. But the Crab Nebula ranks among the few objects in the Milky Way that emit large amounts of *visible-light* synchrotron-emission photons, testimony to the enormous violence of the explosion that produced the nebula.

GREAT PULSARS OF THE MILKY WAY

Pulsars born from collapsing stellar cores have furnished astronomers with a happy situation: their theoretical work rather nicely matches what they observe. Not all supernova sites reveal pulsars; this seems to reflect the fact that some supernovae may produce black holes. Or the radiation from the hot spots of some neutron stars may be beamed in a direction other than our line of sight. In any case, astronomers have now discovered nearly a thousand pulsars in our Milky Way, and several in nearby galaxies. All these pulsars show an impressive regularity in their pulses of radio emission; all are believed to arise from rapidly rotating neutron stars with strong magnetic fields.

The most famous and best-studied pulsar lies within the Crab Nebula. Once a few pulsars had been discovered, and astronomers had begun to speculate (accurately, as it turned out) that pulsars must arise from neutron stars, they concentrated on the center of this supernova remnant, trying to find whether it contains a pulsar. Within a few years, they found that one of the two stars dimly visible within the nebula pulses thirty times per second. Each pulse produces not only radio waves but also visible light and X rays. The Crab Nebula pulsar had been seen—but not as a pulsar!—for more than a century; what had been lacking was the thought (and the equipment) needed to look for, and to find, a star that "blinks" thirty times a second.

For fifteen years after its discovery, the Crab Nebula pulsar stood as the most rapid pulsar known to humanity; most pulsars pulsate a few times per second, or perhaps once every few seconds. But was this so because no more rapidly rotating neutron stars exist, or because astronomers lacked the improved equipment needed to detect more rapid pulsations? In 1982, astronomers at the University of California at Berkeley found a pulsar that spins not thirty but 642 times per second! During the 1980s, six more pulsars spinning more rapidly than a hundred times per second have been detected—creating (for astronomers) a new class of pulsar, the "millisecond pulsars," so named because their rotation rates approach once per millisecond (one thousandth of a second). Do pulsars rotating even more rapidly than this exist, as yet undiscovered? This question can be answered through future efforts.

WHY PULSARS SLOW DOWN—AND SPEED UP

Nothing lasts forever, and the basic trend in pulsars consists of a gradual slowing down, a less rapid rotation as time passes. This slowing arises naturally from the pulsars' emission of electromagnetic waves, the fact that allows us to observe them. Pulsars draw their energy from the rotation of their underlying neutron stars. In order for a pulsar to emit radiation through synchrotron emission, it must draw energy from somewhere: There is no free lunch. In synchrotron emission, the energy in the electromagnetic waves comes from the energy of motion of the charged particles, which have been accelerated to nearly the speed of light by the rapidly rotating magnetic field that sweeps through the surroundings of the neutron star.

The rotating star can accelerate more particles, but in order to do so, it must give up part of the energy of motion in its rotation. As time passes, the neutron star therefore rotates more slowly, so the pulsar pulses more slowly. This phenomenon has been observed over the two decades of pulsar studies. The time interval between pulses from the Crab Nebula object, for example, increases by about thirteen-millionths of a second in every year. With such observations, astronomers believe that they have direct evidence of the slowing of neutron stars' rotation that arises from the radiation emitted by pulsars. As time continues to pass (a universal characteristic of time), the Crab Nebula pulsar and similar objects will pulse more and more slowly, more and more dimly, until some day, a few tens or hundreds of millions of years from now, they will barely qualify as true pulsars.

But some pulsars speed up! In particular, the most rapidly rotating pulsars known until 1989, the type called "millisecond pulsars," those that pulse nearly a thousand times per second, are thought to arise from rather *old* neutron stars. The current theory to explain millisecond pulsars hypothesizes that each of them is a member of a double-star system. In this model, one of the two stars has exploded and left behind a neutron star that produces a pulsar, but the other star has not yet evolved to the point of such collapse and explosion.

In such a situation, the second star may swell up to become a red giant, and material from the red giant may fall onto the surface of its neutron star companion. If the material reaches the star's system from a particular direction, then the neutron star could be set into still more rapid rotation—"spun up," as the theoreticians say—until it rotates not once, or ten or a hundred times, but a thousand times per second!

JANUARY 1989: DID THE PULSAR APPEAR IN SUPERNOVA 1987A?

Theories of stellar explosions predict that many exploding stars should leave behind a rapidly rotating neutron star, the collapsed core of the star that exploded. Since pulsars arise from fast-spinning neutron stars, soon after Supernova 1987A was detected on February 23, 1987, astronomers began to look for regularly spaced "pulses" of radio, visible, and X-ray emission. For two years, their efforts bore no notable fruit. But on the night of January 18, 1989, the situation changed. A team of astronomers from California, Oregon, Canada, and Australia, led by Carl Pennypacker of the Lawrence Berkeley Laboratory (LBL) and John Middleditch of the Los Alamos National Laboratory, had been searching for a pulsar in SN 1987A since March 1987, a few weeks after the discovery of the supernova. Once again members of the team had carried their sensitive, silicon-based photon detectors developed at the LBL, and the incredibly accurate timing equipment to the Cerro Tololo Observatory in Chile, and had mounted it on the 4-meter reflecting telescope. The timing equipment, developed in Pasadena, California, by Jerome Kristian, a scientist on the staff of the Observatories of the Carnegie Institution of Washington, can record data 10,000 times per second. With this apparatus, the team of astronomers apparently made the first detection of the pulsar that Supernova 1987A left behind.

Why was such advanced equipment necessary? A pulsar emits flashes of visible light or radio (often both, along with X rays) at regular intervals of time. However, until astronomers discover a pulsar, they have only the vaguest notion of what the interval between such pulses ought to be. The pulsars detected before 1989 flashed on and off with pulse periods that ranged from a few seconds, for the slowest, up to 682 times per second for the fastest known pulsar. If you want to search for pulsars, you had therefore better plan on searching for pulses that might occur either rather infrequently (on the order of once per second) up to extremely rapidly (many hundreds of times per second).

When the Pennypacker research group planned the observations in Chile that might find visible-light pulsations from Supernova 1987A, they decided to attempt to find flashes of visible light that could occur with potential pulse periods that ranged over a factor of 100,000, from once every few seconds up to 5,000 times per second. (The group decided to allow for the possibility of finding a pulsar flashing even more rapidly than any previously known—in retrospect, a brilliant decision.) The Pennypacker group could search over such a broad range in possible pulse periods because they had designed and built a sensitive "photometer" (light recorder) that could react on time scales far shorter than one five-thousandth of a second. They recorded the data that entered their photometer, focused by one of the giant telescopes in Chile, on magnetic tape and analyzed it on a Cray XMP supercomputer at the Los Alamos National Laboratory in New Mexico.

The Pennypacker group's analysis of the seven hours of observations made on the night of January 18 led to the announcement of the pulsar in early February 1989. However, some doubt remains as to the pulsar's reality, because observations made on the night of January 31, and on many nights thereafer, have failed to detect any pulses from the supernova. However, the majority of astronomers involved believe that the pulses detected on January 18 are real, and that the failure to find detectable pulses on subsequent nights in 1989 occurred because material ejected from the supernova happened to lie directly between ourselves and the pulsar, obscuring our view of the rotating neutron star. Then even though SN 1987A's pulsar continues to emit regularly spaced pulses of light, electrons or other particles along the line of sight would scatter the photons in random directions, "washing out" the pulsation that astronomers seek to detect.

The apparent discovery of a pulsar in Supernova 1987A provided additional confirmation of astrophysical theories about supernovae and the neutron stars they leave behind. Carl Pennypacker compared the discovery to the Belmont Stakes, the third step in horse racing's "triple crown." First the supernova itself, then the neutrino blast, finally the pulsar! The Pennypacker team savored their discovery all the more deeply because of the fact that two weeks earlier, their paper reporting their unsuccessful searches since March 1987 had been rejected for publication in the *Astrophysical Journal Letters.* "No science," the astronomer acting as "referee" had written, meaning that in his opinion, the team's equipment was not sensitive enough to detect a young pulsar; hence the failure to find a pulsar would mean nothing to astronomers.

NEW RECORDS IN PULSAR SPINS

The pulsar found by the Pennypacker group in sweet triumph was no ordinary pulsar. Instead, the Pennypacker group discovered the most rapid pulsar yet known, flashing on and off 1,968.629 times each second. Assuming that these on-and-off pulses arise from "hot spots" close to a rotating neutron star, the conclusion follows that Supernova 1987A left behind a neutron star that spins nearly 2,000 times each second! Since a neutron star has a diameter of about a dozen miles, its circumference equals about 37 miles. Two thousand rotations each second then implies that the surface of the neutron star is being whirled around at 2,000 times thirty-seven miles per second, or 74,000 miles per second—more than one-third of the speed of light!

Spinning at such enormous speeds, the neutron star can barely hold itself together, despite the enormous gravitational force that squeezes the material to densities trillions of times the density of water. Its enormous rotational velocity must have distorted the neutron star into a flattened, oblate spheroid. An observer who could look directly upon the neutron star that Supernova 1987A left behind would see a pumpkinlike object, perhaps twelve miles across at the equator but only four miles from pole to pole, whirling itself nearly to dissolution at 1,968 revolutions per second. As the neutron star rotates, it swings its powerful magnetic field through the regions close by, picking up charged particles and accelerating them to 99.9999 percent of the speed of light and more.

The pulsar in Supernova 1987A is not only the most rapid pulsar ever discovered, but also the first detected pulsar spinning with anything close to this tremendous rotation rate that does *not* arise in a relatively old neutron star. As we have seen, the other "millisecond pulsars" appear in neutron stars that have each been "spun up" by matter falling onto the neutron star from a nearby companion star. The new pulsar's fantastically rapid rotation presumably arises from the fact that Supernova 1987A's pulsar is by far the *youngest* pulsar ever observed by humanity, seen just two years after the explosion of the supernova that left it behind. The previous record holder, the Crab Nebula pulsar, was detected more than nine centuries after the stellar explosion that produced it. As we have seen, pulsars are believed to be born with rapid rotation, and then to slow down gradually as they lose energy through the synchrotron-emission process; some pulsars later "spin up" because of a close companion star. Hence according to theory, SN 1987A's pulsar should be detected in rapid rotation.

But 1,968 spins per second represents something beyond "rapid" rotation! Most theories of neutron star matter in vogue prior to early 1989 predicted that a neutron star rotating this rapidly could not hold itself together, but must break apart into several individual pieces. Once again, nature's laboratory proved capable of rejecting certain theories and supporting others. Since no one can observe matter packed to neutron star densities in a laboratory, astrophysicists must rely on calculations that extrapolate heavily from what we *can* observe. The pulsar in SN 1987A disproved an entire class of theories describing neutron stars, those that rely on a "hard equation of state," and favored those that use a "soft equation of state." These esoteric phrases differentiate between those theories in which the neutrons in a neutron star cannot be packed together quite so closely (a "hard" situation) and those in which such close packing is achievable ("'soft' equations of state"). Only with close packing could a neutron star be small enough not to break apart when spinning nearly two thousand times per second.

WHY DOES THE NEUTRON STAR SPIN SO RAPIDLY?

The rapid pulsation of Supernova 1987A's pulsar, which (most astronomers believe) arises from the equally rapid rotation of the supernova's neutron star remnant, raises a key question: Why did *this* supernova leave behind a neutron star that rotates so rapidly? The answer should be either that the collapsing core that produced the fast-spinning neutron star was itself in relatively rapid rotation, or that the newborn neutron star was "spun up" to a higher rotation rate by matter falling onto its surface.

The former possibility draws on the principle called the "conservation of angular momentum," a law of physics that implies that rotating objects spin more rapidly as they contract. In fact, the rate of rotation increases in proportion to the *square* of the contraction factor, so an object that shrinks to one-tenth its original size should rotate one hundred times more rapidly.

A neutron star, ten miles or so in diameter, forms from the collapse of a core about a thousand times larger. The neutron star should therefore spin not 1,000 but 1 *million*—the square of 1,000—times more rapidly than the stellar core that gave it birth. We think that stellar cores, and stars themselves, typically rotate anywhere from once every few hours (as some stars do) to once each month (as the sun does). If we imagine these cores spinning a million times more rapidly, we reach rotation rates that vary from a hundred times per second (for the cores that were spinning once in a few hours) to once every few seconds. Even the most rapid of these rotation rates fails to match the observed pulsations of SN 1987A's pulsar. We must therefore conclude that this record-setting pulsar arose from a core that rotated once every few minutes, if we hope to produce 1,968 rotations per second through increasing the rotation rate by a mere million times! Just how and why the pre-collapse core should rotate so rapidly may seem mysterious, but the fact that the core itself was contracting, and therefore rotating more rapidly, during millions of years of pre-supernova explosion may provide the explanation.

A second possible explanation of the pulsar's rapid spin, suggested by Stan Woosley and Roger Chevalier, invokes the hypothesis that matter falling back onto the neutron star from the explosion may have made the neutron star spin far more rapidly than would have occurred without such fall-back. In this scenario, some of the supernova's ejected matter would have been pulled back onto the neutron star by its immense gravitational force, and the material would have fallen inwards more in some directions than others. All this would have occurred within a day after the explosion, and the result would have been to make the pulsar spin nearly 2,000 times a second not because it was born that way, but because infalling debris gave its momentum to the neutron star and made it rotate much more rapidly after its first day than when it was born.

The suspected pulsar in SN 1987A should doubtless teach us a few more lessons as it ages. For the first time, astronomers have the chance to observe a pulsar from relatively near birth throughout its career (which will be longer than any individual astronomer's). Observations of the pulsar's evolution will shed new light on the details of how rotating neutron stars create pulses of electromagnetic radiation. Eventually, astronomers hope to detect radio pulses, the signature of most known pulsars, and to study their behavior with time. From such mundane work will arise new edifices of theory concerning stellar collapse and neutron star evolution.

WILL SUPERNOVA 1987A RISE AGAIN?

Astronomers love a good prediction, especially if it turns out to be true. Kenneth Brecher, a supernova expert at Boston University, has made a prediction concerning SN 1987A: In about ten years, just as the third millennium dawns, the supernova should brighten to become visible once again!

Brecher bases his prediction on the recorded behavior of past supernovae. Both SN 1006 and SN 1604 (Kepler's supernova) "reappeared," SN 1006 after ten years, in the year 1016, and SN 1604 after sixty years, in 1664. The evidence of these reappearances rests only in Chinese chronicles, quite appropriate for the supernova of 1006, but a reminder that no European observer recorded the reappearance in 1664.

Both SN 1006 and SN 1604 were Type II supernovae, just as SN 1987A is. Therefore, Brecher notes, even though we know little about *why* a supernova might brighten years after its initial explosion, the odds are good that SN 1987A will do so, perhaps in ten years, perhaps only deep into the twenty-first century. The most likely way for a supernova to rebrighten would arise from the collision of its ejected material, traveling at many thousands of miles per second, with gas and dust relatively close to the exploded star. This matter might consist mainly of material ejected thousands of years earlier from the pre-supernova star, and the high-speed collision of the new ejecta with the old could make the gas glow brightly. Stay tuned to the supernova network, and you will learn whether Brecher was thinking wishfully—or brilliantly.

11

SUPERNOVA 1987A: X RAYS AND GAMMA RAYS

SUPERNOVA 1987A burst upon the public as the first exploding star of general repute since the supernova of 1604—an event which the public had largely forgotten. During the intervening 383 years, and especially during the past few decades, astronomers had used their telescopes to study supernovae in other galaxies. However, because SN 1987A lay closer to the Milky Way than any supernova seen since 1604, astronomers had the chance to observe this exploding star in more detail, and through whole new "channels" of observation.

Chief among these new channels of observation were the neutrinos detected deep underground in Ohio and Japan. These neutrinos were the most significant non-photon emissions from the supernova. And of all the different types of electromagnetic waves—photons—emitted by the exploding star, the most significant were the waves with the highest energy per photon: X rays and gamma rays. Because X-ray and gamma-ray astronomy are still in relative infancy, astronomers have rather meager experience in these fields—especially in comparison with their visible-light expertise. Astronomers still find that supernovae located in galaxies beyond our Local Group are difficult, often impossible, to observe in high-energy photons. But SN 1987A furnished astronomers with a treasure trove of new observations, none of which would have been possible before the advent of satellite-borne astronomical observing platforms. These satellite observatories allowed the detection of types of electromagnetic radiation that can never penetrate the Earth's atmosphere.

THE ERA OF SATELLITE ASTRONOMY

Only with the advent of satellite-borne detectors could astronomers open the new domains of gamma-ray, X-ray, and ultraviolet astronomy. The first such detectors were primitive and prone to malfunction. They "flew" (as rocket scientists still say) during the late 1960s; now, twenty years and several generations of instruments later, dozens of detectors operating at various frequencies are in orbit at any given time. One great change since the 1960s appears in the nomenclature of the instruments, which now bear names such as "Ginga" and "Mir": Instead of remaining a United States monopoly, satellite astrophysics has become internationalized, with the result that Japanese, Soviet, and European experiments typically yield as much (or more) useful data as American satellites.

As Supernova 1987A abundantly demonstrated, we have come a long way in space-borne instruments but still have far to go. What astronomers would dearly love to see orbiting the Earth is a space observatory for each type of photon (save radio photons, which we can observe on Earth): gamma rays, X rays, ultraviolet, infrared, and also visible light. A visible-light space observatory is desirable because atmospheric blurring, even under the clearest skies, allows space observations to provide a better look at the cosmos than any terrestrial observatory can.

No single telescope or photon detector can cover all these frequencies and wavelengths, simply because no single type of material (and therefore no single space-borne detector) can collect, focus, and record all types of photons. (If such material existed, our eyes would doubtless have the capacity to detect infrared and ultraviolet photons, which would have provided an evolutionary advantage.) In order to span the spectrum of all the different types of electromagnetic waves, we therefore require a different type of instrument for each type of photon.

With luck, such specialized telescopes and photon detectors should be in orbit by the end of this century. It would, of course, be sensible if the world's scientific community could collaborate on creating and maintaining such instruments, and indeed some movement toward such a result has occurred, especially among European space agencies. For now, however, a sort of potpourri of space-borne instruments orbits our planet, some with long operating lives, some with short ones; some covering a wide range of photon frequencies and wavelengths, others only a small range; some capable of being serviced in orbit by astronauts, others not; some representing the absolute best technology that we can now produce, others representing a relatively inexpensive and possibly outmoded approach to observing the cosmos from space.

X-RAY AND GAMMA-RAY OBSERVATIONS OF SUPERNOVA 1987A

So it was that crucial observations of high-energy photons from Supernova 1987A came from an old standby of the United States space effort, and demonstrated once again the truth of the adage that good science consists of being ready to observe what you're not ready to observe. In 1980, NASA sent into orbit the Solar Maximum Mission (SMM) satellite, for a well-defined purpose, to study gamma rays emitted by the sun. Like most stars, our sun emits only a tiny fraction of its energy in the form of gamma rays, the highest-energy photons of all. Unusual objects—supernovae among them—emit far larger proportions of their total energy output in gamma rays. However, because all other sources of gamma rays are millions of times more distant than the sun, the strongest observed source of gamma rays remains our own star.

The SMM satellite detects gamma rays with specialized crystals of sodium iodide, which emit flashes of visible light when the high-energy gamma rays strike them. Using this detector system, the SMM aimed to study how the sun's output of gamma rays changed during the course of the "solar cycle," a well-recorded solar variation, with an eleven-year period, during which the number of sunspots (darker regions on the solar surface) waxes and wanes, and the solar magnetic field grows stronger and weaker in concert.

Thanks to solar astronomers' desire to study high-energy photons (gamma rays and the highest-energy X rays), the SMM satellite was already in orbit when SN 1987A exploded. During the 1980s, as the satellite aged, grew cranky, and blew fuses, its orbit also "decayed," that is, grew smaller as the result of friction with the Earth's outermost atmosphere. In 1984, however, the astronauts aboard the Space Shuttle Challenger repaired the satellite and "reboosted" it into a higher orbit, giving it another decade of useful life.

In February 1987, after only slight prodding from supernova-oriented astronomers, NASA arranged for the satellite to study the new potential source of gamma rays. First, the scientists who analyzed the data from the Solar Maximum Mission looked for the record of a large increase in the number of gamma rays recorded at the times when the supernova emitted its first burst of visible light and its first burst of neutrinos. (Even though the SMM satellite was then pointed toward the sun, it would nevertheless have detected a sufficiently strong, sudden emission of gamma rays from another direction, since gamma rays would pass right through the "sides" of such an instrument.) No such gamma-ray burst was found, which fits with the basic model of the supernova, in which all the high-energy photons are trapped *inside* the exploding star, and take several months or years to diffuse outward and escape into space.

After its failure to detect an initial burst of gamma rays from the supernova, the SMM satellite followed a procedure meant to maximize the chance of detecting gamma rays that might arrive from the supernova. In this procedure, the SMM satellite was oriented toward the supernova (this increased the SMM's sensitivity to any gamma rays that the supernova might emit). The SMM satellite was programmed to observe SN 1987A for several weeks' time, and then to add together ("integrate," the astronomers say) all the gamma rays detected during that period. This procedure is standard whenever photons arrive in small numbers, as is the case when observing a supernova in another galaxy; it provides a statistically reasonable method of data analysis—one that avoids the statistical hazards that arise when a very small number of data points (in this case photons) fall within a given interval of analysis.

THE FIRST GAMMA RAYS FROM SUPERNOVA 1987A

In August 1987, the SMM satellite began to record detectable amounts of gamma rays from SN 1987A. These detections were confirmed by balloon-borne detectors launched on flights of only a few hours each, which carried instruments more sensitive than those aboard the SMM satellite. Still more sensitive instruments were launched by balloons from Alice Springs, Australia, in late 1987 and in February, April, and November 1988. These balloon-borne detectors drift at the mercy of the stratospheric winds. Hence stories of recovery of the instrument package take us back to the era when scientists traveled to the most isolated regions of the globe for their discoveries—or, in the modern equivalent, to the rooftops of suburban Rio de Janeiro, to rescue an instrument package before eager souvenir hunters can disassemble it. Fortunately, the astronomers won, and found the record of the high-energy photons.

To "high-energy astrophysicists," the most interesting photons to be observed from SN 1987A are those with energies just in the range where gamma rays, the highest-energy photons, merge into X rays, photons with the next-highest energies (see Figure 7 on page 28). Because astronomers have devoted much attention to studying the supernova at photon energies near the gamma-ray/X-ray boundary, new results reported as "gamma-ray discoveries" and "X-ray discoveries" have, on occasion, turned out to refer to the same observations.

The most interesting photons from SN 1987A turn out to be those with energies near the gamma-ray/X-ray boundary for a good reason. This energy domain includes the photons that arise when nuclei of cobalt-56 decay into other types of nuclei. As we have seen in Chapter 9, astronomers believe that supernovae produce their light through this decay of cobalt-56 during the first few weeks after the initial outburst. The gamma rays from cobalt-56 decay are trapped inside the expanding shell of hot gas, blocked by the various atoms and ions that form this gas. The photons collide with the gas and heat it to a temperature of several thousand degrees, a temperature that causes the gas to radiate photons of visible-light wavelengths.

Eventually, once the expanding shell of gas around the supernova expands to sufficiently large size and a sufficiently low density, some of the gamma rays produced by cobalt-56 can escape directly into space. If this scenario is correct, then within a year or two after the explosion, astronomers should be able to detect some of the gamma rays from the decay of cobalt-56 nuclei. As we know from our laboratory studies of cobalt-56, these gamma rays have certain definite energies. Gamma ray astronomers measure photon energies in units called "keV" (for kilo-electron-volts, or thousands of electron volts). One keV is about one-billionth of the energy of motion of a mosquito. Most of the gamma rays from the decay of cobalt-56 nuclei have energies of 847, or 1238, or 2599 keV.

By August 1987, both the Solar Maximum Mission satellite and the balloon-borne detectors were observing photons from SN 1987A at energies of 847 and 1238 keV. This detection offered both good news and bad news.

On the one hand, the detection of gamma-ray photons with precisely these energies provided direct confirmation of the hypothesis that the supernova explosion had indeed generated a large amount of cobalt-56, since these energies are the signature of cobalt-56 nuclei and of no other types. On the other hand, observations of the visible light from the supernova continued to show a straight-line decline, falling by 50 percent every seventy-seven days. This would be true only if effectively *all* the gamma rays produced by cobalt-56 were being blocked by the gas, and their energy devoted to making visible light by heating the gas. But if the gas blocked all the gamma rays, why did we detect any gamma rays at all?

WERE THE GAMMA RAYS TOO EARLY?

The supernova could not have it both ways: It could not use all its gamma-ray photons to make visible light (by being trapped inside the expanding gas shell, unable to escape and giving up their energy to heating the gas) and also let them escape directly into space so we could detect them. Astronomers had apparently stumbled onto a good thing (the detection of gamma rays) too early. Their favored model of supernova explosions predicted that gamma rays should start to escape directly into space only after a year or more had passed, when the expanding shell of gas from the supernova would have become sufficiently rarefied that it could no longer block all the gamma-ray photons through collisions with atoms and ions. How could this seeming contradiction be resolved?

The answer appears to be to make the model of the supernova explosion a bit more complex, as reality so often is. Imagine that the expanding shell of material blown outward by the supernova, far from being a perfectly symmetric ring of gas, turns out to have a "lumpy" distribution, with "clumps" of greater-than-average density and "holes" of relatively low density. Then, although most of the high-energy photons would be trapped, some could—after a few months—squirt through the "holes" and escape directly into space. These escaping photons would be the ones detected since August 1987, the first direct observational evidence that large numbers of cobalt-56 nuclei exist within the material ejected from a supernova.

FALLING BELOW THE STRAIGHT LINE OF COBALT DECAY

As time passed, what was predicted to occur *did* occur: The supernova's visible-light output declined even *more rapidly* than a straight line (see Figure 18 on page 147). According to the model, which now appears triumphantly verified, the expanding shell of gas could no longer block all the gamma rays and "capture" their energy to heat the gas. Therefore, although cobalt-56 nuclei continued to decay and therefore to produce gamma rays, a declining fraction of those gamma rays ended up heating the gas, and so the visible light from the supernova, which arises from this heating, declined even more rapidly than a straight line with a half life that matches the seventy-seven-day half life of cobalt decay.

X RAYS FROM SUPERNOVA 1987A

Supernova 1987A produced not only gamma-ray and visible-light photons but also X-ray photons that proved of immense interest to astronomers. A photon's energy is its essence: Since a photon has no mass, effectively all that it possesses is its energy, its ability to make an impact when it strikes an object. X-ray photons each have enough energy to penetrate human flesh (hence their use in medicine), and now that astronomers can send spacecraft above the atmosphere, they can detect and analyze these photons, which arise not in ordinary stars but in noteworthy, often explosively special, cosmic objects.

Unlike gamma rays, X rays do not arise from the decay of nuclei such as cobalt-56. Instead, X-ray photons, like those of visible light, are typically produced by a hot gas simply because it *is* hot. However, while temperatures of thousands of degrees will produce visible light, the temperature must rise toward a million degrees to produce significant amounts of X rays.

In supernova explosions, gamma rays come from the fundamental process in the exploding star: Nuclei fused during the explosion decay into other types of nuclei, producing gamma-ray photons as they do so. In contrast, the X-ray photons arise later on, once the gamma rays, blocked by the gas, have given their energy to the shell of material exploding from the supernova. This energy heats the gas to temperatures of hundreds of thousands, or even millions of degrees. At these temperatures, the gas "glows," but in X rays, not in visible light, as it would at temperatures of a few thousand degrees. X-ray observations of an exploding star therefore do not provide us with a direct look at the photons made when nuclei decay, as gamma-ray observations do. But the X rays observed from SN 1987A do furnish an opportunity to look at what is going on within the shell of gas as it expands.

These X-ray observations could not be made by NASA's X-ray-observing satellites: There really aren't any. NASA hopes eventually to launch the world's best X-ray observational platform, the Advanced X-ray Astrophysics Facility or AXAF satellite. For now, that project remains in doubt as to funding. As a result, X rays from SN 1987A have been detected (since August 1987) and studied by Soviet and Japanese astronomers, the former with the "Kvant" ("quantum") module aboard their space station "Mir" ("world" or "peace"), the latter with the "Ginga" ("milky way") satellite.

One important difficulty does hinder any attempt to observe X rays from SN 1987A: X-ray telescopes have an extremely "fuzzy" view of the universe, for they cannot observe the sky with anything like the crisp resolving power of visible-light telescopes. This problem arises from the relative infancy of our X-ray detector technology, but for now we are stuck with a blurry view of the universe in X rays. As a result, our "X-ray eyes," the X-ray detecting satellites in orbit above the atmosphere, have a view of the cosmos even worse than that of a highly nearsighted man, who has been deprived of his eyeglasses and squints at the world, barely able to distinguish the trees from the buildings.

For Supernova 1987A, an additional problem arose in making

X-ray observations. As ill luck would have it, one of the most intense X-ray sources already known to astronomers lies relatively close to the supernova in the sky, at a distance of only about twice the full moon's diameter. Thus the X-ray satellites are in the position of a squinty-eyed, nearsighted gentleman trying to find a dim lightbulb in almost the same direction as a spotlight. The intense source of X rays, called LMC X-1, was the first X-ray source detected in the Large Magellanic Cloud, the home galaxy of Supernova 1987A. LMC X-1 is probably a neutron star onto which matter is falling from a companion star, heating up and emitting X rays as it rushes toward the intense gravitational field of the neutron star. For visible-light observations, the separation between LMC X-1 and SN 1987A would be entirely adequate to distinguish the two easily, but for X-ray observations this is not the case. Instead, when we try to study X rays from the fainter source (in this case the supernova), stray X rays from the more intense source may enter the detector, causing confusion as to just what we are observing. For this reason, all the X-ray observations must be taken with caution as representing the true state of affairs.

The Japanese scientists who operate the "Ginga" satellite, fully aware of this problem, believe that they have accomplished the task of distinguishing between X rays coming from LMC X-1 and those from Supernova 1987A. Ginga's instruments are sensitive though less so than Kvant's to X rays of almost all energies and frequencies, and the scientists began to look for X rays from the supernova on February 25, 1987, the day after the supernova was detected.

For four months, Ginga found no X rays, but then, in late June of 1987, a detectable amount of X-ray photons appeared for the first time. This flux of X rays steadily increased in intensity until the end of August. Since August 1987, the amount of X rays has remained roughly constant. This corresponds to the predictions made by most models of how X rays can leak from within the expanding shell of gas after being initially trapped inside it, when the density of gas within the shell is so high that the gas blocks all the X rays.

WHAT DO THE X RAYS TELL US?

When the Japanese analyzed the number of X-ray photons of each energy, they found that SN 1987A shows a far larger ratio of the number of higher-energy X-ray photons to the number of lower-energy X-ray photons than other X-ray sources do. Astronomers say that the X rays from SN 1987A are "harder" (show comparatively more high-energy photons) than other X-ray sources. Nearly every other source that produces X rays produces fewer high-energy X rays than low-energy X rays. In contrast, SN 1987A produced as many high-energy as low-energy X-ray photons. To an astronomer, this is a striking anomaly, one that makes SN 1987A unique in the catalog of cosmic X-ray sources. We still have little idea of why this supernova produced proportionately more high-energy X rays, and theoreticians will be working on this problem for some time to come.

The Soviet X-ray observations of Supernova 1987A generally confirmed the Japanese results, at least for the high-energy X rays. But in observing low-energy X-ray photons, the Japanese found an interesting result that the Soviets apparently missed. Although low-energy X rays were first detected from SN 1987A in late July 1987, at the same time as the high-energy X rays, they decreased in intensity by a factor of four during late September 1987. This decrease seems highly unlikely to arise from a decrease in the number of cobalt-56 nuclei at that time, since the high-energy X rays continued to emerge in the same numbers. If the Japanese observations are correct (and most astronomers believe that they are), it seems likely that different processes—as yet to be completely specified—may be responsible for the high-energy and low-energy X rays.

FUTURE PROSPECTS FOR OBSERVING SUPERNOVA 1987A

Astronomers who care about supernovae look forward to the next few years with eagerness, since they will be able to discover further details about SN 1987A that have general relevance to stellar explosions. By careful study of the gamma-ray emission from the expanding shell of gas, they can hope to determine whether or not the shell includes material created in the deep interior of the star, just above the core, and mixed in with more outlying material when the star exploded. Deep-lying material would be richer in heavy nuclei than the outermost layers, and these nuclei will produce gamma rays at different wavelengths than lighter elements will. The question of how much mixing occurs in a supernova's shell bears directly on the question of just which nuclei, and in what amounts, add to the interstellar mulch when a star explodes. This in turn determines how rapidly, and in what elements, the cosmos enriches itself in heavy elements.

To astronomers who study pulsars, the most fascinating question about SN 1987A remains the one that we discussed in Chapter 10: Did the supernova create a pulsar, a rapidly rotating neutron star that produces beams of intense synchrotron emission? If it did—if the observations made in January 1989 indeed recorded this pulsar—then we know that we have glimpsed the visible light produced by this snychrotron process. The excellent agreement between the actual visible light output from the supernova and the theoretical light curve based on the decay of cobalt-56 nuclei does show that as of now, a pulsar (if one exists) cannot be contributing significantly to the *total* visible light from the explosion. If it were, the visible light would not be declining so smoothly, with a half life of seventy-seven days, just what we expect from the decay of cobalt-56.

Calculations show that if a neutron star does exist within SN 1987A, and if this neutron star's magnetic field has the same strength as the neutron star that produces the pulsar in the Crab Nebula, then the neutron star in SN 1987A cannot be rotating more rapidly than about fifty times per second. More rapid rotation of the neutron star would produce more visible light by the synchrotron process, since the neutron star's magnetic field would be slung through space at a higher rate, thereby accelerating charged particles in its vicinity to higher velocities. If the rotation rate exceeded fifty times per second, the neutron star would be a pulsar producing so much visible-light emission that we would have already detected this emission, superimposed on the emission from the expanding shell of the explosion. The total emission from the supernova—pulsar plus expanding shell of gas—would then violate the observed straight-line decline of the visible-light output.

The failure to detect a pulsar would pose a puzzle, since from what we know about the evolution of pulsars, we would expect a newborn pulsar to rotate at least fifty times per second, and we would also expect that any neutron star should have about the same strength of its magnetic field as the one in the Crab Nebula. However, it is possible that a relatively slowly rotating pulsar (rotating fewer than fifty times per second!) was born in SN 1987A, and will yet make its presence known to us, and it is also possible that the neutron star producing the apparent fast spinning-pulsar is relatively weak in its magnetic field. Meanwhile, astronomers enjoy the tension: Either the pulsar will be confined, gladdening the hearts of those who predicted that the supernova should leave behind a pulsar, or it will not, giving further work to the theoretical astronomers, who must explain not merely what we ought to see, but what we do see, in the cosmos that surrounds us.

TO FLARE AGAIN: THE SHOCK WAVE MEETS THE EJECTA

As SN 1987A's shock wave moves outward, accelerating the matter it meets away from the supernova, it will eventually encounter the material that the pre-supernova star expelled into space during its red giant period, when its outer atmosphere swelled into an enormous, highly rarefied envelope that slowly evaporated. This material, puffed into space tens of thousands of years before the star exploded, should lie at distances ranging from a fraction of a light-year up to several light-years from the site of the supernova.

Since the supernova's shock wave now travels outward at about 20,000 miles per second—one-tenth the speed of light—it will plow into the circumstellar material in two to twenty years, more or less, after the explosion. The shock wave will then heat this material to high temperatures, causing it to shine in X rays and in radio waves. (The shock may also make the star reappear in visible light, as we discussed in Chapter 10). Analysis of this radiation will allow astronomers to determine which elements were produced by the pre-supernova star—information that will aid them in their quest to understand the final stages of a star's evolution before it collapses and explodes.

Supernova 1987A will therefore have a claim to astronomers' attention for at least the remainder of the twentieth century. With their appetites thus whetted, astronomers of the twenty-first century may well fulfill the hope of observing a supernova in our own galaxy, not so close as to be dangerous, but much closer than SN 1987A, perhaps at one-tenth the distance of the greatest supernova of this century. When and if they do so, the instruments and techniques used for the supernova in the Large Magellanic Cloud—suitably improved, of course—will serve as the fundamental means for collecting the data and testing the theories that allow scientific progress. Until then, we shall have to make do with Supernova 1987A, long anticipated yet completely unexpected, an explosion from past ages that helps to expand the frontiers of astronomical knowledge.

12

SUPERNOVAE AS PROBES OF THE UNIVERSE

THE PREVIOUS chapters have discussed the main aspects of supernovae: Exploding stars not only brighten the night skies, but in addition have provided the universe with nearly all of its elements other than hydrogen and helium. If we seek a true appreciation of this cosmic mulching by supernovae, we must examine how the universe made hydrogen and helium before supernova explosions made the rest of the elements. Once we see how supernovae fit into the universal scheme of things, we can admire how supernovae may yet provide the answer to a great riddle of the cosmos: will the universe expand forever?

MULCHING THE UNIVERSE THROUGH SUPERNOVA EXPLOSIONS

What we familiarly call "matter" on Earth consists of atoms. An atom has one or more electrons in orbit around a nucleus made of protons and neutrons, each of which has more than 1,800 times the mass of an electron. If you could enlarge an atom by 10 trillion times, you could visualize the protons and neutrons in the atomic nucleus as ball bearings, half an inch across (see Figure 9 on page 48). Then the electrons would each weigh a tiny fraction of an ounce, and would orbit the nucleus at distances of nearly a mile! We consist of atoms scaled like this, but 10 trillion times smaller: there is more empty space in the universe than one realizes at first.

Every chemical element—each universal collection of basically similar atoms—is defined by the number of *protons* in the nucleus of the atom. Hydrogen has one proton per nucleus, helium has two; carbon has six protons per nucleus, nitrogen seven, oxygen eight, and so on. Thus the number of protons, which always equals the number of electrons, plays the dominant role in determining what type of an atom we have.

Without stellar evolution, the universe would consist of the two simplest elements, hydrogen and helium. This would have made chemistry and nuclear physics far simpler than they are, though we would not be here to admire their simplicity. But over the immense eons since stars began to form, 10 billion years or more ago, in the greatest "pollution" the universe has ever known, exploding stars have produced the nuclei with more protons than hydrogen or helium. These nuclei could later form the centers of atoms, since they captured electrons to orbit around them.

In this way, supernovae made the element carbon; they made the oxygen, nitrogen, silicon, sulfur, phosphorus, fluorine, and chlorine; made the metals such as iron, copper, magnesium, aluminum, titanium, and zinc; made the noble gases neon, argon, krypton, xenon, and radon; made the "rare earths" lanthanum, samarium, gadolinium, and holmium; made all the silver and gold; and made the radioactive elements uranium and thorium. As supernovae exploded from time to time throughout the past ten billion years, they "mulched" galaxies such as the Milky Way with nuclei other than hydrogen and helium. In this way, supernovae produced the matter that now forms the Earth, the air, and life itself. In the most egocentric terms, each of us consists of a small lump of matter processed through exploding stars and later incorporated into ourselves. To be frank, red giant stars' expanded outer shells have contributed much of the carbon, nitrogen, and oxygen, but all the heavier elements come from exploded stars.

Picture, as an example, what will happen to the parts of Supernova 1987A that were blasted into space. These outer layers of the now-defunct star contain plenty of hydrogen and helium, which does little to alter the universe, but they also contain significant amounts of nuclei such as carbon, nitrogen, oxygen, neon, and iron, and smaller amounts of nuclei such as fluorine, manganese, sulfur, and chlorine.

The most abundant elements in the supernova's "ejecta" can be identified by the fact that they absorb, or emit, only certain wavelengths of light. Hence when astronomers spread the visible light—and the ultraviolet—emitted by the supernova into its different colors, they can spot the "fingerprint" of a given element. This fingerprint consists of the relatively large or small amounts of light of a particular color, caused by the presence of a particular element. The more abundant such an element is, the stronger its spectral fingerprint will become. Astronomers have now identified about two dozen different elements in the spectrum of SN 1987A. They are confident that the supernova made some amount of all, or nearly all, the ninety-two elements that occur naturally in nature. In short, Supernova 1987A, a typical supernova, has apparently made its invididual contribution to the universe in *all* the elements.

But how do these star-blasted elements ever find themselves in stars and planets? The answer is slowly, and partially. Most of the nuclei made by SN 1987A and blasted outward into space are moving at speeds of a few thousand miles per second, testimony to the force of the explosion that produced them. These nuclei will slow down as they encounter the gas and dust already floating in interstellar space. Eventually, the nuclei will capture electrons, some of which are also floating among the stars (electrons, too, are blasted into space when stars explode), and will form atoms. Nuclei made in SN 1987A will, after a few tens of millions of years, mingle with the existing interstellar medium, adding their small contribution to its composition. In a hundred million years or so, no particular trace of SN 1987A will remain; instead, its contribution will be well mixed with the overall interstellar loam in the Large Magellanic Cloud.

From this loam—the clouds of interstellar gas and dust—new stars continually form, as we described in Chapter 6. Somewhere between a few hundred million and a few billion years from now, it is likely that new stars will be born that incorporate part of SN 1987A, as well as parts of thousands of other supernovae that exploded in the Large Magellanic Cloud. Although the *light* (and the neutrinos) produced by SN 1987A escaped easily from the supernova's home galaxy, almost all of the nuclei will remain forever in the Large Magellanic Cloud. The sole exception to this rule concerns some of the highest-energy "cosmic-ray" particles, the tiny fraction of nuclei that the supernova blasted into space with almost the speed of light. But the star-building stuff stays in its own galaxy. SN 1987A will never help to form stars in the Milky Way; for that, we must look to the supernovae that explode among our own 400 billion stars.

Supernova explosions progressively add a bit more of the "heavy" elements—those with nuclei more complex than hydrogen and helium—to interstellar gas and dust. Therefore, as stars form over billions of years, they form from matter that has become progressively richer in these heavy elements. When our sun formed, 4.6 billion years ago, about one percent of its mass consisted of nuclei other than hydrogen and helium. If the sun were to form today, this fraction might be up to one and one-quarter, or even to one and one-half of a percent of the total mass. This is progress on a galactic scale: Supernovae have continued to make the Milky Way richer in heavy elements.

WHERE IS THE *EARTH'S* HYDROGEN AND HELIUM?

But our Earth consists not of 1 percent, but of nearly 100 percent of elements heavier than hydrogen and helium. How did this occur? The answer is that the sun began to shine. According to astronomers' best current theories, each of the sun's planets formed at about the same time that the sun did, as a sub-condensation within the rotating pancake of matter that shrank to form the solar system. Most of the matter ended up at the center of the condensation, to form the sun; smaller clumps aggregated at different distances from the sun to form the planets and their satellites; and a host of still smaller condensations became the comets, each with the mass of a small mountain, that orbit the sun far beyond all the planets.

Once the sun began to produce light and heat, while the clumps that became the planets were in their final stages of aggregation, it affected its "protoplanets," the planets in formation. Clumps relatively far from the sun received little of its warmth, and therefore underwent little change in their chemical composition. These clumps became the giant planets, Jupiter, Saturn, Uranus, and Neptune, which consist to this day primarily of hydrogen and helium, like the sun and the rest of the universe. But the inner protoplanets, the ones that became Mercury, Venus,

Earth, and Mars, lay closer to the sun, and therefore grew far warmer once the sun began to shine. In fact, these four clumps of matter lost most of their hydrogen and helium through evaporation, simply because hydrogen and helium, the lightest atoms, escape into space most easily of all.

Hence when we look around the inner solar system for hydrogen and helium, we find almost none, save in the sun itself, where the enormous amount of self-gravitation retains even these lightest elements, despite the high temperatures within the sun. But on Mercury, Venus, Earth, and Mars, you will look in vain for hydrogen and helium, except for rather small amounts of water on Earth and Mars, and trace amounts of helium trapped in pockets inside the inner planets. The inner planets are the husks of might-have-been Jupiters, had the sun not warmed these protoplanets to the point that most of their original matter evaporated into space, leaving behind only the heavy elements that we know and love. For these elements we can thank the stars that exploded, billions of years ago, long before the sun and its planets were born. To say that these explosions have enriched our lives is to err on the side of understatement: They made us what we are today.

SUPERNOVAE AS PROBES OF THE EXPANDING UNIVERSE

To astronomers as well as the general public, the mulching of the universe by supernova explosions would by itself rank supernovae high among cosmically significant events. But exploding stars possess another aspect that holds promise to astronomers, not for what supernovae *do* to the universe but what they may *tell* us about it. Supernovae may hold the key to resolving a basic mystery of the universe: Will the universe expand forever, or will it someday cease its expansion to begin a universal contraction? In order to appreciate the importance of the answer to this question, it helps to comprehend the question itself, and to understand what it means to say that the universe is expanding.

THE EXPANDING UNIVERSE

To paraphrase Mark Twain, everyone knows that the universe is expanding, but few of us know what to do about it. Astronomers, of course, know what to do: They seek to study the present and the past in order to learn the future of the universe.

Astronomers have concluded that the universe is expanding ever since 1929, when Edwin Hubble discovered that except for the galaxies in our own Local Group, all of the galaxies in the universe are receding from our own, with speeds that increase in direct proportion to the galaxies' distances from us. A galaxy cluster twice as far from us as another is receding twice as rapidly, and one three times more distant has a recession velocity three times larger. This holds true for all the galaxies that we observe, in all directions looking outward from our own Milky Way.

Hubble's discovery, since improved and enlarged, hinges upon a relationship between the *distances* to clusters of galaxies and the *velocities* at which those clusters are receding from the Milky Way. The measurement of recession velocities uses the "Doppler effect," the change in the frequency and wavelength of light waves from a source in motion toward or away from that observer. The Doppler effect for *sound* waves furnishes an everyday physics experiment, beloved by teachers the world over: Go outdoors and listen to the lonesome wail of a steam locomotive, or (to modernize the experiment) the piercing shriek of an ambulance. You will notice that the sound you hear has a higher frequency (that is, a larger number of vibrations per second) when the source of the sound is approaching you, and a lower frequency when it passes you and recedes into the distance.

The same effect for light waves furnishes astronomers with their basic, indispensible tool for measuring the velocities of faraway galaxies. Astronomers have grown familiar over the years with the distribution of colors—the *spectrum*—in the light that they observe from stars, and from galaxies made of billions of stars. Because most stars have a characteristic pattern of more and less light in the various colors of the spectrum, astronomers have grown familiar with this pattern when they observe stars and galaxies throughout the sky.

If astronomers find such a familiar pattern of the colors in an object's spectrum, but one displaced in frequency, so that the entire pattern shows (for example) a lessening of one percent in all the frequencies that form the spectrum, they conclude that the Doppler effect has worked its everyday magic. The source must be receding from us at 1 percent of the speed of light; if the frequencies were all reduced by (say) 3 percent, the recession velocity would be 3 percent of the light speed. This effect depends only on the velocity of the source with respect to the observer, and not at all, for example, on the distance to the object. In this way, astronomers have measured the speeds at which clusters of galaxies are receding from the Milky Way, and have found that the farther those clusters are from us, the more rapidly they are receding.

WHERE IS THE CENTER OF THE UNIVERSE?

We might conclude from the fact that galaxies' recession velocities are proportional to their distances from the Milky Way that our Milky Way galaxy must be the center of the expanding universe. Although this conclusion receives easy emotional verification, astronomers have long since forced themselves to reject any hypothesis that our planet, our star, or our galaxy occupies a special, central position in the cosmos.

For purely philosophical reasons, astronomers impose on their speculations a principle of cosmic modesty, which they name the "cosmological principle": Our view of the cosmos is a representative one, so that any observer, anywhere in the universe, sees the same sort of things that we do. If we use this principle (remember that it is only an hypothesis), we must conclude that *any* observer in *any* galaxy will see what we see: Galaxies are receding from *that* observer, and at speeds that are proportional to the galaxies' distances from *that* observer. And if this is so, then *every* observer sees galaxies receding from *that* observer, so the entire universe must be in a state of expansion, everywhere, with galaxy clusters behaving something like a vast swarm of bees, all moving away from one other.

This conclusion rests on the assumption that we *do* have a representative view of the universe, and this assumption could prove incorrect. It might turn out, for example, that galaxies are receding from one another in one part of the universe, but somewhere else—far beyond our limits of vision—galaxies are approaching one another. Astronomers have adopted the cosmological principle simply because any alternative assumption produces a more complex (and unknowable) universe, and scientists like to see where the simplest hypothesis will lead. If the principle is correct, then the entire universe must be expanding.

How can the entire universe expand *with no center* to its expansion? The best model that astronomers can provide suggests that we imagine the entire universe as the *surface* of an expanding balloon. If non-expanding dots on the balloon represent galaxies, then when we blow up the balloon, each dot "sees" all the other dots moving away, and at speeds that increase in proportion to the distances of the other dots. All that you need to do to make this model "work" in your mind is to imagine that *only* the skin of the balloon exists, so that there is no inside nor outside (these are simply useful in seeing the balloon clearly). The balloon's skin represents all of three-dimensional space. Therefore light travels only around the balloon's surface, and the dots "see" one another only along that surface.

THE BIG BANG

The universal expansion, by definition, implies that clusters of galaxies are growing more distant from one another, everywhere. This in turn implies that in the past, galaxies used to be closer to each other. If we imagine that we have a movie of the history of the universe, and run that movie backwards, we reach a point about 15 billion years in the past when matter in the universe had near-infinite density. This moment in time, called the "big bang," represents the beginning of the universe, at least in its present state. Even our best theories tell us little about how the universe ever came to have a "big bang," but the observed behavior of the universe *today* strongly suggests that the big bang *did* occur some 15 billion years ago.

GREAT MOMENTS IN THE EARLY UNIVERSE

During the first microseconds after the big bang, the universe was immensely *hot*. The enormous temperatures of all the matter in the universe, billions upon billions of degrees, arose simply because matter was concentrated at an enormous density—everywhere. Squeezing gas into a smaller volume makes it hotter, as you know if you have ever studied a diesel engine, which ignites the fuel in its cylinders simply by compressing the fuel–air mixture into a smaller volume. The early universe was one hell of a diesel engine. As the universe expanded, the density of matter decreased, and like the expansion cycle of a diesel, this expansion cooled the entire universe.

The temperature of a gas measures the average energy of motion per particle in the gas. During the first seconds after the big bang, the temperature in the universe exceeded a billion degress, so the entire universe formed a roiling cosmic caldron, in which every particle had far more energy of motion than any particle does in the center of a star today. As a result of their enormous energies, particles collided with immense fury, countless times per second, and from these collisions new particles and their antiparticles emerged, which in turn collided to make new particles, and so on endlessly—so long as the universe was extremely hot.

As the universe expanded, it cooled. A crucial moment arose at a time about half an hour after the big bang. Before then, each particle had sufficient energy of motion that when it collided with another particle, new types of particles were likely to be made. But as the universe expanded and cooled, the average energy of motion of each particle decreased, until at a time about half an hour after the big bang, collisions between particles no longer produced new types of particles. Instead, collisions simply bounced the colliding particles off one another in random directions. Hence the first half hour after the big bang was the time when the basic mixture of particle types was established in the universe. Since that time, one half hour after the universe began its expansion, the universe has undergone only local, not universal changes in its composition.

Those local changes have mostly been supernova explosions. Inside stars, the temperature and density have been temporarily high enough to fuse helium into more complex nuclei, as we have examined in detail. Using the basic fuel in the universe, hydrogen nuclei, stars that exploded have made all the elements except hydrogen and helium. If this were all that supernovae did, it would be enough. As we have seen, however, exploding stars do more: They probably produce the cosmic rays that allow evolution to proceed. And there is another aspect to supernova explosions that deserves our attention, not for what supernovae have done but for what we can do with them. What we can do—we hope!—is to use supernovae to obtain reliable distance measurements for faraway galaxies.

ESTIMATING THE DISTANCES TO GALAXIES

Astronomers who study the expanding universe have one vast regret. (Many of them have more, but this one is common to all.) Although they can measure galaxies' recession velocities with quite respectable accuracy, thanks to the Doppler effect, astronomers unfortunately have no technique nearly so neat to measure the *distances* to galaxies. Even the Andromeda galaxy, one of the galaxies closest to the Milky Way, has a distance of two million light years—half a million times more distant than the closest star to the sun. For such distant objects, astronomers must fall back on *estimates,* most of which rely on a simple rule of physics: The apparent brightness of any object decreases in proportion to the *square* of the object's distance from the observer. We use this rule instinctively, though we are probably not aware of the "inverse-square" behavior of apparent brightnesses; after all, our ancestors never hunted for prey that revealed itself with mounted headlights!

To astronomers, the inverse-square brightness law is second nature. As a result, astronomers approach the problem of estimating distances to a faraway object with one simple hope, often dashed: If they can find some object—a star, a gas cloud, a stellar nursery—in the faraway object that they *think* is identical (more or less) to a similar, closer object whose distance they know, then they have an easy job. Simply compare the relative apparent brightness of the two objects, and the inverse-square brightness law will reveal the ratio of the objects' distances. Suppose, for example, that we think two objects are identical, and the more distant object has an apparent brightness equal to one one-hundredth of the closer object's apparent brightness. Then the more distant object must be the square root of one hundred, or ten times, farther away. This method rests upon a knowledge of the distance to the *closer* object and upon the assumption that the two objects are identical, or so nearly so that we can justifiably assume that they emit the same amount of energy each second in the form of light waves. Otherwise we might be comparing, in cosmic terms, the headlights of a semitrailer with the dim lamp on a bicycle, thinking that we had identical objects—an exercise sure to yield an incorrect estimate of the distance to the farther objects.

Undaunted by these twin problems, astronomers have proceeded to estimate the distances to thousands of galaxies by comparing the apparent brightness of objects that they believe to be nearly identical. Their results have deepened their confidence in the relationship that Hubble found: More distant galaxies are indeed receding from us more rapidly, and at speeds proportional to their distances. As we observe more and more distant galaxies, we can use the galaxies *themselves* as the objects for comparison. Since we have established the distances to the nearer galaxies by observing objects within those galaxies, we can now find the ratio of distances for a relatively nearby and a more distant galaxy of the same shape, by comparing the apparent brightnesses of the entire galaxies. If the fainter galaxy appears, for example, one ten-thousandth as bright as the brighter galaxy, it must be one hundred times more distant. This method has proven triumphant—up to a point. That point has a distance from us of five to ten billion light years, the crucial sort of distance for predicting whether or not the universe will expand forever.

THE FUTURE OF THE EXPANDING UNIVERSE

Generations of astronomers have come to accept as commonplace the conclusion that we live in an expanding universe, a conclusion that rests (firmly, in the astronomers' view) on the notion that what *we* see forms a representative slice of reality. The big question then becomes, will the universe expand forever? To this we have as yet no definitive answer, for one crucial reason: We have tremendous difficulty in estimating the *distances* to faraway galaxies.

EXTRAPOLATING FROM THE PAST INTO THE FUTURE

When we look at galaxies billions of light-years away, we see the galaxies not as they are, but as they were half the age of the universe ago. This is just the point: By studying such faraway galaxies, and by determining the past history of the universe, we can hope to determine the future of the universal expansion. Comparison of the way that the universe *was* expanding, many billions of years ago, with the way that it is expanding *now,* should allow astronomers to extrapolate into the future, and to determine whether the expansion will ever cease.

But if faraway galaxies' enormous distances allow us to look far back into the past—and they do—they also raise a problem. Astronomers know that galaxies must have changed significantly over billions of years, as the stars within them aged, and some exploded. They therefore have little confidence in the apparent-brightness method for estimating the distances to galaxies that we see as they were 6 or 8 or 10 billion years ago. Astronomers need another, more accurate method to estimate distances like these. They need supernovae.

SUPERNOVAE AS DISTANCE ESTIMATORS

How do supernovae provide hope for resolving astronomers' distance-measurement dilemma? First, supernova explosions are so intrinsically luminous that astronomers can detect supernovae even in galaxies many billions of light-years from the Milky Way. It would be great news if supernovae provided a perfect "standard candle"; great, that is, if all supernovae had the same absolute or intrinsic luminosity at the time of their maximum light output. In that case, we could simply measure the maximum apparent brightness of a supernova seen in a distant galaxy, and could determine how much farther that galaxy must be from us than a galaxy in which a supernova of greater maximum apparent brightness was observed. Unfortunately, what we have found out about supernovae in relatively nearby galaxies—those whose distances have been reliably estimated by other means—dashes this hope on the rocks of reality.

Supernovae, even when sorted out by Type I or Type II show too much variation among individual explosions, some reaching much greater luminosities at maximum light than others do. Supernova 1987A furnishes a good example: Apparently because it became blue before it blew, SN 1987A had a peak luminosity much less than most Type II supernovae do. Supernova explosions therefore cannot provide W. C. Fields' elusive spondulix, the magic "standard candle," visible at enormous distances. Must we abandon hope for using supernovae as an accurate means of estimating the distances to the galaxies in which they appear? Not at all.

Several astronomers, most prominently Robert Wagoner of Stanford University, have noted that supernova explosions provide another means to estimate galaxies' distances. This method, which Bob Kirshner has used several times, relies not on the maximum luminosity (light output) from the supernova, but on a different aspect of the explosion: the likelihood that (at least on the average) a star's explosion sends material outward at the same velocity *in all directions*.

Let us assume that this is true (and it seems reasonable to adopt it as a working hypothesis). If it is true, then a supernova explosion offers the chance to combine two separate observations that will reveal the supernova's distance from us. One of these observations is the *speed* at which material is ejected from the supernova toward us, which we can measure by using the Doppler effect, the change in the wavelength and frequency of light caused by motion toward or away from an observer. The Doppler effect does not depend on an object's *distance,* but only on the relative speed with which the object is approaching us, or receding from us, along our line of sight to the object. For example, the Doppler effect reveals that material ejected from Supernova 1987A is now approaching us at 20,000 miles per second (one-tenth of the speed of light), and we measure this speed independently of our knowledge (or lack of knowledge) concerning the distance to SN 1987A.

The second observation to be made involves the expansion of the exploding star in directions *perpendicular* to the line from ourselves to the supernova. That is, we want to measure the motion of the expanding shell of gas not in the directions toward or away from us (which we can do by using the Doppler effect), but in all directions across our line of sight.

In an ideal world, we could do this by measuring how rapidly the shell of material ejected by the supernova appears to grow larger. Suppose that we could measure this expansion by making careful observations of a supernova at different times—say, at one year, two years, and five years—after the detection of the supernova. The apparent diameter of the supernova would be measured as a tiny angle on the sky—a fraction of a second of arc. One second of arc equals 1/3600 of a degree, and it takes 360 degrees to circle the entire sky.

The observed rate of expansion depends on the supernova's distance from us. For any actual rate of expansion, measured in miles per second, the supernova's apparent diameter, measured in seconds of arc, will increase more slowly if the supernova is farther from us. *Greater* distances imply a *lesser* rate of increase in the supernova's apparent diameter because more distant objects appear smaller to us, and so too do their increases in size.

If we could observe the supernova's apparent rate of expansion, measured in seconds of arc per year, then we could combine this observation with the velocity of the expansion, which we obtain in miles per second via the Doppler effect. This combination would give us the distance to the supernova, since we know how rapidly the diameter of a supernova at any given distance should appear to increase if the material in the supernova's shell expands at a certain number of miles per second. If the supernova shell expands at 20,000 miles per second and the supernova's distance equals 1 million light-years, we would see its diameter appear to increase by one-tenth of a second of arc in five years. But if the supernova were 2 million light-years away, its diameter would appear to increase only half as rapidly, by one-twentieth of a second in five years.

But a serious problem exists with this method. If a supernova explodes in a galaxy outside our Milky Way, we cannot hope to measure the increase in the supernova's apparent diameter directly. A supernova in another galaxy, especially in a relatively distant galaxy, is just too far away for us to detect it as anything but an intense point of light, even five or ten years after the explosion. Our best telescopes simply can't measure angles as small as one-twentieth of a second of arc, because of the blurring that our atmosphere produces. It might therefore appear that although the method under consideration makes perfect sense in theory, in prarctice we can't use it.

Nevertheless, we *can* determine the diameter of the supernova, using *indirect* means. We can do this because we think we understand the basic physics of the explosion—how the expanding shell of gas produces its light as the atoms heated by the shock wave produce photons. This understanding allows us to measure the *temperature* within the expanding shell from careful observations of the supernova's spectrum of light—the same observations that reveal the speed of the explosion through the Doppler effect.

Once we know the temperature of the expanding shell of gas, we can *calculate* the apparent diameter of the supernova from our observations of the supernova's apparent brightness. The supernova's *absolute brightness* (brightness as seen from a standard distance) depends on two quantities: the temperature in the expanding shell of gas and the total surface area of that shell. The supernova's *apparent* brightness (brightness that we observe) depends on the temperature of the expanding shell of gas, the total surface area of the shell, and the supernova's distance from us. But now a happy fact of geometry helps with the problem. Both the *apparent area* that an object covers on the sky and its *apparent brightness* decrease with distance, and in the same way: If you moved the moon to twice its present distance, it would appear one-quarter as bright as it does now, and it would appear to cover one-quarter as much area on the sky as it does now.

Because both the apparent brightness and the apparent area of an object on the sky decrease with the distance in the same way, we obtain an important simplification in using supernovae as distance estimators. If one supernova is four times farther from us than another, identical supernova, it will have one-sixteenth of the closer supernova's apparent surface area—and one-sixteenth of its apparent brightness. This means that the supernova's apparent brightness *per unit of surface area* will remain unchanged, because both the apparent brightness and the amount of surface area have decreased in the same proportion.

With this simplification—the fact that the apparent brightness per unit of surface area does not depend on the distance to the supernova—astronomers can derive a formula that relates a su-

---

FIGURE 21. The Veil Nebula in the constellation Cygnus consists of material from a star that exploded tens of thousands of years ago. This gas and dust will eventually merge with the general mass of interstellar material, perhaps to be incorporated into a new generation of stars. (National Optical Astronomy Observatories)

pernova's distance to two factors, the speed of its expanding outer layers and the temperature in those layers, which we can measure by observing the different colors of light that the supernova emits. Astronomers are now in a position to use this method to obtain increasingly accurate estimates of the distances to faraway galaxies in which supernova explosions occur. Eventually, this may become *the* best method for measuring distances to galaxies that are several billion light years away from us.

On December 11, 1989, NASA plans to launch the long-awaited Hubble Space Telescope, an automated reflecting telescope above the atmosphere, capable of observing the universe in ultraviolet as well as in visible light. The Space Telescope's avoidance of atmospheric blurring, and its ability to observe at ultraviolet wavelengths, should make it the premier instrument for the technique that we have described above. The Space Telescope should therefore provide us with distance estimates of increased accuracy for those galaxies in which we observe supernova explosions.

If this method yields distance estimates for faraway galaxies as accurately as astronomers hope they will, supernovae may yet resolve the question, will the universe expand forever? This would allow astronomers to add another stripe to their supernova coat of arms. Supernovae, which made the heavy elements, made our planet, made ourselves, and made the cosmic rays that help drive evolution, would also have provided the key to the mystery of what future lies in store for the universe. From the cosmic catastrophes of stellar collapse would come not only the elements essential for our lives but also information crucial for our comprehension of the universe. From death, rebirth; from disaster, understanding: These are the lessons we learn from exploding stars such as Supernova 1987A.

GLOSSARY

Absolute temperature scale—Temperature measured on a scale that uses the same units as the Centigrade (Celsius) scale but sets the zero point at absolute zero, so that water freezes at 273.16 degrees and boils at 373.16 degrees.

Absolute zero—The lowest point in temperature, at which all motion ceases (except for certain quantum-mechanical effects). Absolute zero occurs at −273.16 degrees Centigrade or −459.67 Fahrenheit.

Absorption—The removal of photons of a particular wavelength and frequency, usually as the result of the photons' interaction with atoms or molecules.

Absorption line—A limited region of the spectrum of photons within which the intensity of the radiation falls below that of the neighboring spectral regions.

Acceleration—A change in an object's speed, or its direction of motion, or both.

Angular size—The part of a circle over which an object appears to extend. Angular size is measured in degrees (360 in a circle), minutes of arc, and seconds of arc.

Antielectron—The antiparticle of an electron (also called a positron), with a mass equal to an electron's mass and one unit of positive electric charge.

Antineutrino—The antiparticle to a neutrino, with the same mass as a neutrino (either zero or an extremely small mass) and no electric charge.

Antiparticle—The complement of a particle, with the same mass but opposite sign of electric charge. If brought together with its corresponding particle, mutual annihilation results.

Atom—The smallest unit of an element, consisting of a nucleus with one or more protons and none or more neutrons, surrounded by one or more electrons in orbit around the nucleus. The number of electrons always equals the number of protons in the nucleus.

Atomic nucleus—The center of an atom, containing nearly all the atom's mass. Each nucleus includes one or more protons and none or more neutrons.

Atomic number—The number of protons in an atom.

Big bang—The primeval explosion, approximately 15 billion years ago, that started the universe in its present state of expansion.

Black hole—An object with such enormous gravitational force at its surface that neither matter nor electromagnetic radiation (including visible light) can escape from it.

Bolometric light curve—The light curve (the record of changes in brightness plotted on a graph) that includes electromagnetic radiation of all wavelengths and frequencies.

Cassiopeia A—A source of radio waves, the brightest radio source in the constellation Cassiopeia, apparently the remnant of a supernova that exploded about 300 years ago.

Centrigrade (Celsius) temperature scale—The scale of temperature that registers the freezing point of water at 0 degrees and the boiling point of water at 100 degrees.

Chandrasekhar mass limit—The maximum mass that any white dwarf star can have, an upper limit equal to 1.4 times the sun's mass.

Cobalt-56—An isotope of cobalt with 28 protons and 28 neutrons per nucleus that is unstable and is subject to radioactive decay. The half life of this decay is 77.1 days.

Cosmic rays—Particles moving at nearly the speed of light through interstellar space. They are thought to arise, at least in part, in supernova explosions. Most cosmic-ray particles are electrons, protons, or helium nuclei.

Crab Nebula—The remnant of the star observed to explode in the year 1054 in the constellation Taurus. The Crab Nebula was the first cosmic object found to produce visible light through the synchrotron process, by which a pulsar that is believed to have arisen from the collapsed core of the exploding star generates visible light and other types of electromagnetic radiation.

Degenerate matter—Matter in which the exclusion principle plays an important role in determining how the matter can move.

Degree of arc—One three-hundred-sixtieth of a full circle.

Density—The amount of mass per unit of volume.

Deuterium—An isotope of hydrogen that has one proton and one neutron per nucleus (hydrogen-2).

Doppler effect—The apparent change in the frequency and wavelength of electromagnetic waves (or of other types of waves) that reach an observer from a source of waves that is approaching, or receding from, that observer.

Doppler shift—The amount of the change in frequency or wavelength caused by the Doppler effect.

Electromagnetic forces—One of the four basic types of forces, acting between particles with electric charge, either as a repulsive force (between particles with the same sign of electric charge) or as an attractive force (between particles with opposite signs of electric charge, i.e., plus and minus).

Electromagnetic radiation—See electromagnetic waves.

Electromagnetic waves—Streams of photons carrying energy from a source. The photons are characterized on the basis of their frequency and wavelength as gamma rays, X rays, ultraviolet, visible light, infrared, or radio waves.

Electron—An elementary particle with one unit of negative electric charge and a mass of 9.1×10^{-28} gram, one of the three basic particle types in an atom.

Electron volt—A unit of energy used by physicists, equal to 1.602×10^{-12} erg.

Element—The set of all atoms that have the same number of protons in the atomic nucleus.

Elementary particle—A fundamental particle of nature, one that is indivisible into smaller particles.

Elliptical galaxy—A galaxy with an ellipsoidal distribution of stars, hence one whose shape appears elliptical on a photograph.

Energy—In scientific terminology, the capacity to do work, that is, the capacity to exert a given amount of force over a specified distance.

Energy of mass—The energy contained within a given amount of mass simply by virtue of the existence of the mass, equal to the amount of mass times the square of the speed of light.

Energy of motion—Energy associated with the motion of an object, also called kinetic energy. A particle with mass m and velocity v has an energy of motion equal to one-half m times the square of v.

Erg—A unit used to measure energy. A mass of two grams moving at a speed of one centimeter per second has an energy of motion equal to one erg.

Exclusion principle—The rule of nature that no two particles of the same type can have almost the same location and almost the same velocity. This rule applies to certain types of elementary particles, most notably to protons, neutrons, and electrons.

Exponential decay—Change of one particle type into other types that is characterized by the decay, within a given amount of time, of a particular fraction of all the particles present that have not previously decayed.

Force—The capacity to cause a physical change in an object, usually manifested as an acceleration of the object in the direction in which the force is applied.

Frequency—The number of times that a photon vibrates each second, measured in units of cycles per second ("hertz").

Galaxy—A large group of stars, usually along with some gas and dust, held together by the mutual gravitational forces among the stars.

Galaxy cluster—A group of galaxies, held together by the galaxies' mutual gravitational attraction, typically containing a few dozen to a few thousand individual member galaxies.

Gamma rays—Photons with the shortest wavelengths, largest energies, and highest frequencies, usually defined as photons with energies greater than a few hundred thousand electron volts.

Gravitational forces—One of the four basic types of forces, always "attractive." For any two particles with mass, the amount of gravitational force varies in proportion to the product of the particles' masses, divided by the square of the distance between their centers.

Half life—The time required for half the nuclei in a sample of a particular type of unstable nuclei to decay into other types of particles.

Helium—The second-lightest and second most abundant element, whose nuclei each contain two protons.

Hubble's Law—The summary of the expansion of the universe, which states that the velocities of galaxy clusters as they move away from us equals a constant times the clusters' distances from us.

Hubble Space Telescope—The automated reflecting telescope with a 94-inch mirror, to be launched by NASA in 1990 into an orbit above the Earth's atmosphere.

Hydrogen—The simplest and most abundant of the elements, with a nucleus of one proton plus zero, one, or two neutrons.

IMB detector—An enormous tank of pure water, half a mile underground, operated by Brookhaven National Laboratory, the University of Michigan, and the University of California, Irvine. The device can function at a sensitive detector of neutrinos.

Infrared—Electromagnetic radiation consisting of photons with slightly longer wavelengths and slightly lower frequencies than those of visible light.

Interstellar absorption—The blockage of starlight by dust particles in interstellar space.

Interstellar matter—Matter spread among the stars in a galaxy such as our own Milky Way, consisting primarily of gas (mostly hydrogen and helium), along with other atoms, some molecules, and larger dust particles.

Ion—An atom that has lost one or more of its electrons.

Ionization—The process of making an ion.

Iron-56—An isotope of iron, its most abundant form, with twenty-six protons and thirty neutrons in each atomic nucleus.

Irregular galaxy—A galaxy whose shape appears neither spiral nor elliptical.

IUE satellite—NASA's International Ultraviolet Explorer satellite, in geosynchronous orbit at an altitude of 22,300 miles, capable of observing celestial objects' ultraviolet radiation.

Kamiokande detector—The detector at Kamioka, Japan, buried deep underground, similar to the IMB detector, and likewise capable of detecting neutrinos from cosmic objects.

Kelvin temperature scale—See absolute temperature scale.

Kepler's supernova—The supernova observed to appear in the constellation Ophiuchus in the year 1604, extensively studied and recorded by Johannes Kepler.

Kinetic energy—Energy associated with the motion of an object, also called energy of motion.

Large Magellanic Cloud—The Milky Way's largest satellite galaxy, an irregular galaxy about 160,000 light-years from the sun.

Light curve—The record of changes in the brightness of an object over time.

Light echo—The phenomenon by which a source of light may appear to have rings of light surrounding it, the result of the source's light reflecting from dust particles relatively close to it, and taking somewhat longer paths to reach the observer.

Light-year—The distance that light travels in one year, equal to about 6 trillion miles.

Local Group—The small cluster of about twenty galaxies to which our Milky Way, its satellites, and the Andromeda galaxy belong.

Luminosity—The total energy-per-second emitted by an object as electromagnetic radiation.

Magnetic field—An invisible field of force in space, created by a magnet or electric current, that changes the trajectories of electrically charged particles.

Magnitude—A measure of the relative brightness of objects, on a scale in which larger numbers indicate fainter objects, and each unit of magnitude signifies a decrease by a factor of 2.512 in brightness.

Main-sequence star—A star in the prime of life, which fuses protons into helium nuclei as a steady rate.

Mass—A measure of the amount of matter contained in an object, often determined by measuring the resistance of the object to being accelerated by a given amount of force.

Milky Way—The spiral galaxy of which the sun is a member, whose central regions appear as a band of light or "milky way" on the sky as seen from Earth.

Minute of arc—One-sixtieth of a degree of arc.

Mystery spot—The source of light seen about one-twentieth of a second of arc from Supernova 1987A, which may or may not have arisen from an actual object.

Nebula—A diffuse mass of interstellar gas and dust, often lit from within by young, hot stars that have recently formed within it.

Neutrino—A particle with no electric charge and with no mass (or an extremely small mass), characteristically emitted or absorbed in particle interactions governed by weak forces.

Neutron—An elementary particle (actually made of three particles called quarks) with a mass of 1.6747×10^{-24} gram and no electric charge, stable when part of an atomic nucleus but subject, when not part of any nucleus, to decay with a half-life of eighteen minutes into a proton, an electron, and an antineutrino.

Neutron star—A tremendously dense object, typically about a dozen miles in diameter, formed from the core of a collapsed star, in which almost all of the particles have formed neutrons, and which is supported against further collapse by the exclusion principle.

Nickel-56—An isotope of nickel, consisting of nuclei which each have twenty-seven protons and twenty-nine neutrons, unstable and subject to decay with a half life of six days.

Nova—A star that shows a sudden increase in brightness, but much less of an increase than the tremendous jump in brightness of a supernova.

Nuclear fusion—The joining together of nuclei by strong forces, typically reducing the total energy of mass and increasing the total energy of motion.

Nucleus—The central region of an atom, composed of one or more protons and none or more neutrons.

Photon—An elementary particle with no mass and no electric charge, which forms electromagnetic radiation and always travels at the speed of light, i.e., 186,000 miles per second, in empty space.

Planetary nebula—A shell of gas surrounding an aging star, heated by the radiation from the star, and which has been previously ejected from the star itself.

Proton—An elementary particle (actually made of three quarks) with a mass of 1.6724×10^{-24} gram and one unit of positive electric charge, one of the basic constituents of an atomic nucleus.

Pulsar—An object that emits pulses of electromagnetic radiation at regularly spaced intervals of time, thought to arise from a rotating neutron star.

Quark—An elementary particle that comes in several varieties and that, taken three at a time, forms protons and neutrons.

Radiation—See electromagnetic radiation.

Radioactive decay—The process by which unstable nuclei change into other types of nuclei, which often includes the emission of gamma-ray photons, neutrinos, and antineutrinos, through the influence of weak forces.

Radio waves—Electromagnetic radiation with the longest wavelengths and lowest frequencies.

Red giant star—A star that has ended its prime of life phase, has begun to exhaust its supply of protons for nuclear fusion, and has a contracted core and an expanded, rarefied outer envelope of gas that has cooled to only a few thousand degrees Fahrenheit, and therefore shines in red light rather than in yellow or blue light.

Red supergiant star—A red giant star with a particularly large size and a particularly high luminosity.

Second of arc—One-sixtieth of a minute of arc.

Shock wave—A disturbance within a gas characterized by a sudden increase in the density and pressure of the gas, and which travels through the gas at the speed of sound.

Sk −69° 202—The star that exploded as Supernova 1987A, one of the stars in the catalog of blue stars compiled by Nicholas Sanduleak.

SN designation—The astronomical denotation of supernovae, in which the letters SN are followed by the year and by a letter indicating the rank order of discovery within that year.

SN 1987A—The supernova detected in explosion in the large Magellanic Cloud on February 23, 1987.

Speckle interferometry—A technique used to obtain a clear view of an object despite the blurring introduced by the Earth's atmosphere.

Spectroscopy—The observation and analysis of the spectra of light from celestial objects.

Spectrum (plural, spectra)—The distribution of photons in frequency and wavelength, often shown as the number of photons with each particular frequency and wavelength.

Spiral galaxy—A galaxy characterized by a flattened disk of stars, within which the youngest, brightest stars are distributed in a spiral-arm pattern.

Standard candle—A source of radiation of known luminosity, which can be used in the determination of distances by comparing the apparent brightnesses of such sources.

Star—A self-luminous mass of gas held together by self-gravitation, in which the kinetic energy released through nuclear fusion balances the star's tendency to contract.

Strong forces—One of the four basic types of forces, always attractive, which act only among certain types of elementary particles (in particular, among protons and neutrons), which has an effect only for distances of 10^{-13} centimeter or less, and which holds together the protons and neutrons in an atomic nucleus.

Supernova—An exploding star, visible for weeks or months even at tremendous distances, because of its enormous luminosity.

Supernova remnant—The exploded outer layers of a star that became a supernova; also, the collapsed core that may result from a supernova explosion.

Synchrotron emission—Electromagnetic waves emitted when electrically charged particles, moving at nearly the speed of light, change either their speed or their direction (or both) while moving in the presence of a magnetic field.

Synchrotron radiation—See synchrotron emission.

Tarantula Nebula—The enormous nebula, part of the Large Magellanic Cloud, close to the site of the explosion of Supernova 1987A.

Temperature—The measure of the average kinetic energy of random motion within a group of particles. On the absolute temperature scale, the temperature is directly proportional to the average kinetic energy per particle.

Thermonuclear fusion—See nuclear fusion.

Tycho's supernova—The supernova observed to explode in the constellation Cassiopeia in the year 1572, first detected by the Danish astronomer Tycho Brahe.

Type I supernovae—Supernovae believed to originate from white dwarfs that acquire large amounts of new material from a companion star.

Type II supernovae—Supernovae that arise from the collapse of a star's core, which bounces outward slightly to start a shock wave that blasts the star's outer layers into space.

Ultraviolet—Electromagnetic radiation with frequencies somewhat greater, and wavelengths somewhat less, than those of visible light.

Unstable nucleus—An atomic nucleus that changes ("decays") into another type of nucleus, typically requiring anywhere from a fraction of a second up to thousands of years to do so.

Visible light—Electromagnetic radiation with wavelengths and frequencies that can be detected by human eyes.

Wavelength—The distance between successive wave crests or wave troughs; for photons, the distance that a photon travels while it vibrates once.

Weak forces—One of the four basic types of forces, acting only among certain types of elementary particles and over distances of 10^{-13} centimeter or less, responsible for the decay of certain types of elementary particles into other types.

Weak reaction—An interaction among elementary particles in which weak forces are important.

White dwarf—A star that has fused helium nuclei into carbon nuclei before becoming so dense in its interior that the exclusion principle supports the star against further contraction. The star continues to radiate stored energy but generates no new energy through nuclear fusion.

Work—In physics, the measure of energy expended by a force, measured by the product of the amount of force applied to an object and the distance over which the force is applied as the object moves.

X rays—Electromagnetic radiation with frequencies greater than those of ultraviolet but less than those of gamma rays.

FURTHER READING

BOOKS

Isaac Asimov, *The Exploding Stars* (New York: Dutton, 1985).
David Clark, *Superstars* (New York: McGraw-Hill, 1984).
George Greenstein, *Frozen Star* (New York: Freundlich Books, 1983).
Laurence Marschall, *The Supernova Story* (New York: Plenum Publishing Company, 1988).
Simon Mitton, *The Crab Nebula* (New York: Scribner's, 1978).
Paul Murdin and Leslie Murdin, *Supernovae* (London: Cambridge University Press, 1985).

MAGAZINE ARTICLES

Hans Bethe and Gerald Brown, "How a Supernova Explodes," *Scientific American* (May 1985), p. 60.
Adam Burrows, "The Birth of Neutron Stars and Black Holes," *Physics Today* (September 1987), p. 28.
Paul Gorenstein and Wallace Tucker, "Supernova Remnants," *Scientific American* (July 1971), p. 74.
David Helfand, "Supernovae: Creative Cataclysms in the Galaxy," in *The Universe,* edited by A. Fraknoi (New York: Bantam Books, 1987).
Ronald Kahn, "Desperately Seeking Supernovae," *Sky and Telescope* (June 1987), p. 594.
Robert Kirshner, "Supernovas in Other Galaxies," *Scientific American* (December 1976), p. 88.
Ronald Schorn, "Happy Birthday, Supernova!" *Sky and Telescope* (February 1988), p. 134.
Frederick Seward, Paul Gorenstein, and Wallace Tucker, "Young Supernova Remnants," *Scientific American* (August 1985), p. 88.
F. Richard Stephenson and David Clark, "Historical Supernovas," *Scientific American* (June 1976), p. 100.
J. Craig Wheeler and Robert Harkness, "Helium-Rich Supernovae," *Scientific American* (November 1987), p. 240.

INDEX